T0139886

SEMA SIMAI Springer Series

ICIAM 2019 SEMA SIMAI Springer Series

Volume 6

This sub-series of the SEMA SIMAI Springer Series aims to publish some of the most relevant results presented at the ICIAM 2019 conference held in Valencia in July 2019.

The sub-series is managed by an independent Editorial Board, and will include peer-reviewed content only, including the Invited Speakers volume as well as books resulting from mini-symposia and collateral workshops.

The series is aimed at providing useful reference material to academic and researchers at an international level.

More information about this subseries at http://www.springer.com/series/16499

María Isabel Asensio • Albert Oliver • José Sarrate
Editors

Applied Mathematics for Environmental Problems

Editors
María Isabel Asensio
Department of Applied Mathematics
University of Salamanca
Salamanca, Spain

Albert Oliver
University Institute SIANI
University of Las Palmas de Gran Canaria
Las Palmas de Gran Canaria, Spain

José Sarrate
Department of Civil and Environmental Engineering
Universitat Politecnica de Catalunya
Barcelona, Spain

ISSN 2199-3041 ISSN 2199-305X (electronic)
SEMA SIMAI Springer Series
ISSN 2662-7183 ISSN 2662-7191 (electronic)
ICIAM 2019 SEMA SIMAI Springer Series
ISBN 978-3-030-61797-4 ISBN 978-3-030-61795-0 (eBook)
https://doi.org/10.1007/978-3-030-61795-0

This Springer imprint is published by the registered company Springer Nature Switzerland AG.
The registered company address is: Gewerbestrasse 11, 6330 Cham, Switzerland

Preface

The profound study of nature is the most fertile source
of mathematical discoveries.
Joseph Fourier

This book contains four papers from the contributions presented at the Applied Mathematics for Environmental Problems minisymposium during the International Congress on Industrial and Applied Mathematics (ICIAM) held in July 15–19, 2019, in Valencia, Spain.

Climate change, air and water pollution, deforestation, and soil degradation are some of the world's biggest environmental problems that humanity needs to face. They are major challenges with large economic and social impacts that, if not addressed, will increase in the near future.

How can mathematics and numerical modelling help us wrestle with these environmental problems? Mathematics is the common language of science and engineering for developing models that help understand nature and life. Numerical modelling allows scientists and engineers solve these models using computers and helps them understand, quantify, predict, and manage the consequences of these problems and devise potential solutions.

The contributions presented during this minisymposium cover several environmental models and their numerical and computational treatment. These models are based on partial differential equations and solved using different numerical methods, combined with efficient computational techniques to provide useful forecasting tools in decision-making and warning.

The first two papers are devoted to modelling wildfire, one of the environmental problems that climate change is worsening. Wildfire spread models can be an efficient aid to combat this growing problem, not only in wildfire management, but also in risk mapping, reforestation policies, and the major issue of alerts and evacuation plans. One of the papers deals with a simplified physical wildfire spread model, based on partial differential equations solved with finite element methods and integrated into a Geographical Information System to provide a useful and efficient tool. The other paper focuses on one of the causes of the unpredictable behavior of wildfire, fire-spotting, through a statistical approach.

The third paper addresses low-level wind shear (LLWS) that represents one of the most relevant hazards during aircraft takeoff and landing. Specifically, it presents an experimental wind shear alert system based on predicting wind velocities obtained from the Harmonie-Arome model.

The final paper deals with the environmental impact of oil reservoirs. It presents high-order hybridizable discontinuous Galerkin (HDG) formulation combined with high-order diagonally implicit Runge-Kutta schemes to solve one-phase and two-phase flow problems through porous media.

We warmly thank all the speakers and participants for their contributions and discussions during and after the minisymposium. In addition, we would like to thank the anonymous reviewers of the papers for helping us to improve the quality of this volume. Finally, we acknowledge SEMA SIMAI Springer Series for their interest in publishing these contributions. We hope that the papers included in this book will be of interest to the SEMA SIMAI Springer Series community and will contribute to the development of new mathematical tools for environmental problems.

Salamanca, Spain — María Isabel Asensio
Las Palmas de Gran Canaria, Spain — Albert Oliver
Barcelona, Spain — José Sarrate

Contents

About the Editors

María Isabel Asensio received her PhD in mathematics from the University of Salamanca, where she is associate professor in the Department of Applied Mathematics and principal investigator of the Numerical Simulations and Scientific Computation Research Group. Her main research interests concern mathematical modelling of environmental problems with emphasis on forest fire spread modelling. She has coauthored many articles in international journals and refereed contributions in international congresses.

Albert Oliver received his PhD from the Universitat Politècnica de Catalunya and is currently an assistant professor at the University of Las Palmas de Gran Canaria. His research interest is the application of the finite element method in environmental problems, and particularly in the simulation of air quality and wind fields at the microscale level; he also works on the generation of tetrahedral adapted meshes for these problems.

José Sarrate obtained his PhD in applied sciences from the Universitat Politècnica de Catalunya, Barcelona, where he is associate professor in the Department of Civil and Environmental Engineering in Barcelona. His research interest is the numerical modelling of problems governed by partial differential equations in applied sciences and engineering. He has worked in more than 20 research projects funded by public and private institutions that lead to several contributions in indexed journals and international conferences.

About the Editors

PhyFire: An Online GIS-Integrated Wildfire Spread Simulation Tool Based on a Semiphysical Model

M. I. Asensio, L. Ferragut, D. Álvarez, P. Laiz, J. M. Cascón, D. Prieto, and G. Pagnini

Abstract The PhyFire simplified physical wildfire spread model developed by the research group on Numerical Simulation and Scientific Computation at the University of Salamanca has been integrated into an online GIS interface in order to facilitate its use, automate the data input process, thereby reducing error and improving efficiency, and upgrade the graphical display of simulation results. The main features of the PhyFire model are presented: model equations, numerical solution and GIS integration. A description is provided of new advances in the PhyFire model related to the addition of random phenomena, such as fire-spotting. A real wildfire simulation with fire-spotting is also presented.

1 Introduction

Environmental issues are a global priority, with wildfires being a classic example, as they are becoming more serious as climate and climate change strongly influence their activity [2, 16], making catastrophic forest fires more likely every year: California, the Amazon, Australia... the list is endless.

M. I. Asensio (✉) · L. Ferragut · D. Álvarez · P. Laiz · J. M. Cascón
University Institute of Fundamental Physics and Mathematics, University of Salamanca, Salamanca, Spain
e-mail: mas@usal.es; ferragut@usal.es; daalle@usal.es; plaiz@usal.es; casbar@usal.es

D. Prieto
Cartographic and Terrain Engineering Department, University of Salamanca, Ávila, Spain
e-mail: dpriher@usal.es

G. Pagnini
BCAM-Basque Centre for Applied Mathematics, Bilbao, Spain

Ikerbasque-Basque Foundation for Science, Bilbao, Spain
e-mail: gpagnini@bcamath.org

M. I. Asensio et al. (eds.), *Applied Mathematics for Environmental Problems*,
SEMA SIMAI Springer Series 6, https://doi.org/10.1007/978-3-030-61795-0_1

Predicting the behaviour of complex environmental systems is an undeniably useful tool for reducing their negative effects, which in some cases are devastating. Mathematical modelling plays a fundamental role. Wildfire modelling is a tool for understanding and predicting fire behaviour, and it can ultimately assist in decision-making in preventing and fighting fires. Improved spatial information technology, advanced computational capabilities and the development of communication technology have exponentially increased the efficiency and applicability of wildfire modelling. The combined use of technological advances and quasi-empirical models has resulted in comprehensive tools for the prediction of wildfire spread, such as FARSITE [15], Prometheus [27], FIREMAP [28], or SFIRE [18], among the most widely applied.

Nevertheless, the rapid increase in computing power allows more complex models to be a real option, so research is focusing on physical-based models. Within this framework, the authors have improved their simplified physical wildfire spread model PhyFire [4, 23], combining it with their high-definition windfield model HDWind [12] to provide an efficient online tool to predict forest fire spread in Spain through the url: http://sinumcc.usal.es.

The PhyFire model is a single-phase simplified 2D physical model based on the fundamental physics of combustion and fire spread, which considers convection and radiation as the main heat transfer mechanisms, taking into account the heat lost by natural convection, the effect of the flame tilt caused by wind or slope, and the influence of fuel moisture content and fuel type. The resulting partial differential equations (PDEs) are solved using efficient numerical and computational tools to obtain a software with levels of efficiency comparable to empirical models.

As wind is one of the factors that most influences wildfire spread, PhyFire can be coupled with the HDWind model. This model provides a high-resolution wind field that adjusts specific wind measurements to local characteristics, such as the slope, the roughness of the terrain and the temperature gradients on the surface. The HDWind model is based on an asymptotic approximation of the Navier-Stokes equations, resulting in a mass-consistent vertical diffusion model capable of providing a 3D wind field by solving only 2D linear equations.

The development of these models, and especially the software necessary to transform them into accessible, efficient and useful tools, is based on multidisciplinary collaboration, leading to the development of the online geographic information system (GIS) tool presented here, which allows open access to PhyFire and HDWind models, and which considers the incorporation of new models and tools for several environmental issues.

The outline of the paper is as follows. Section 2 describes the PhyFire model equations, focussing on the latest improvement added to the model, namely, the fire-spotting module. Section 3 briefly summarizes the architecture of the online GIS interface developed to transfer the developed models to the community. Section 4 adds a real case of a wildfire that involved fire-spotting. We finish with the conclusions and acknowledgements.

2 PhyFire Model

PhyFire is a simplified 2D wildfire spread model developed by the Research Group on Numerical Simulation and Scientific Computation (SINUMCC) at the University of Salamanca in Spain. It is based on principles of energy and mass conservation. Although it is a 2D model, it takes into account important 3D factors: the model considers the energy lost in the vertical direction, and the two most important heat transfer mechanisms in wildfires: radiation and convection. Radiation from the flames above the surface takes into account the effect of wind and slope over flame tilt. It is a single-phase model: only the solid phase is considered, with the gaseous phase being parameterized through flame temperature and flame height in a radiation term. The convective term is critical, as wind is one of the most influential factors in a fire spread. Fuel moisture content is also considered by using an enthalpy multivalued operator. The current versions of the PhyFire model equations are detailed in the following subsection, together with the input model variables, model parameters and output variables. A specific subsection is devoted to the introduction of random phenomena, specifically fire-spotting. Finally, certain details about the numerical resolution of the model equations are summarized. This numerical solution is implemented in C++ using the Neptuno++ library, a Finite Element library developed by L. Ferragut [7].

The first version of the PhyFire model was a simple single-phase 2D physical model first published in [8], based on the energy and mass conservation equations, considering only convection and diffusion. Radiation was first incorporated with a local radiation term in [4]; the radiation term was subsequently modified by a non-local radiation term [9], which allows modelling the radiation from the flame above the fuel layer, thereby considering the effect of wind and slope over flame tilt. The influence of moisture content and heat absorption by pyrolysis through the multivalued operator representing the enthalpy has been reported in [10]. The latest improvements have been made after a global sensitivity analysis of the model on both experimental and real examples [22] and [6]. The most important enhancement was the introduction of a flame length sub-model in [6]. Finally, the model's latest advancement, which is described here, is the addition of a random term for the simulation of fire-spotting. Improvements have also been made to the efficiency of numerical resolution algorithms [14], the adjustment of parameters by simulating real fires [13], and the model's accessibility and usability through the development of a GIS tool [23].

2.1 Current PhyFire Model Equations

Let S the surface where the fire occurs, with height given by a known function h, whereby

$$\begin{aligned} S : d &\longmapsto \mathbb{R}^3 \\ (x, y) &\longmapsto (x, y, h(x, y)), \end{aligned}$$

where $d = [0, l_x] \times [0, l_y] \subset \mathbb{R}^2$ is the rectangle representing the projection of surface S. The size of the surface S and the study time interval $(0, t_{\max})$ are selected so that the fire does not reach the boundary in time $t_{\max}$, to assume homogeneous boundary conditions to complete the set of PDEs that governs the PhyFire model. The latest version of these non-dimensional PDEs are as follows:

$$\partial_t e + \beta \mathbf{v} \cdot \nabla e + \alpha u = r + q \quad \text{in } S \times (0, t_{\max}), \tag{1}$$

$$e \in G(u) \quad \text{in } S \times (0, t_{\max}), \tag{2}$$

$$\partial_t c = -g(u)c \quad \text{in } S \times (0, t_{\max}), \tag{3}$$

where the unknowns are the following variables defined in $S \times (0, t_{\max})$: the dimensionless enthalpy e, the dimensionless solid fuel temperature u, and the solid fuel mass fraction c. The relationship between these dimensionless variables and the corresponding physical quantities is given in Table 1.

These physical quantities are the enthalpy E, the solid fuel temperature T, and the fuel load M. Their relationship depends on a reference temperature T_∞, the heat capacity of the solid fuel C and the maximum solid fuel load M_0. The reference temperature T_∞ is related to the ambient temperature, measured far enough from the fire front to ensure that $T \geq T_\infty$, whereby $u \geq 0$. The other weather data that feed the PhyFire model are wind direction and intensity given by $\mathbf{v}$, and ambient humidity, which affects the fuel moisture content M_v. The heat capacity of the solid fuel C, and maximum solid fuel load M_0, both depend on fuel type, as do all the other input model variables in Table 2, as detailed below.

Finally, the PhyFire model depends on three model parameters listed in Table 3. These three parameteres are related to the heat transfer terms in (1): α in the natural convection term αu, β in the convective term $\beta \mathbf{v} \cdot \nabla e$, and a in the radiation term r, and they will be further explained later on in this section.

Table 1 Dimensionless unknowns and related physical quantities

Physical variable	Symbol	Units	Dimensionless variable
Enthalpy	E	$J\,m^{-2}$	$e = E/MCT_\infty$
Solid fuel temperature	T	K	$u = (T - T_\infty)/T_\infty$
Solid fuel load	M	$kg\,m^{-2}$	$c = M/M_0$

Table 2 Fuel-type-dependent input variables

Input variable	Symbol	Units
Heat capacity	C	$J\,K^{-1}\,kg^{-1}$
Maximum fuel load	M_0	$kg\,m^{-2}$
Moisture content	M_v	kg water/kg fuel
Flame temperature	T_f	K
Pyrolysis temperature	T_p	K
Combustion half-life	$t_{1/2}$	s
Flame length independent factor	F_H	m
Flame length wind correction factor	F_v	$m^{1/2}s^{1/2}$
Flame length slope correction factor	F_s	–

Table 3 Model parameters

Parameter	Symbol	Units
Natural convection coefficient	H	$Js^{-1}m^{-2}K^{-1}$
Convective term factor	β	–
Mean absorption coefficient	a	m^{-1}

We complete the problem with homogeneous Dirichlet boundary conditions and the following initial conditions,

$$u(x, y, 0) = u_0(x, y) \quad \text{in } S, \tag{4}$$

$$c(x, y, 0) = c_0(x, y) \quad \text{in } S. \tag{5}$$

representing the source of the fire and initial fuel distribution, including any possible firebreaks.

We should point out that the data for orography (function $h(x, y)$ defining S), solid fuel distribution (initial value of M), and fuel type distribution (in order to spatially define the fuel-type-dependent input variables) are obtained from the cartography generated specifically for each case. This process has been automated by developing the online GIS interface and the necessary cartographic database. Details of this process are explained in Sect. 3.

There now follows an explanation of all the terms in the PhyFire model equations and their relationship with the aforementioned input variables and parameters.

We shall begin by explaining the multivalued Eq. (2) that models the influence of moisture content and depends on the fuel moisture content M_v and solid fuel pyrolysis temperature T_p. The dimensionless enthalpy e is thus an element of the multivalued maximal monotone operator G defined by

$$G(u) = \begin{cases} u & \text{if} \quad u < u_v, \\ [u_v\,,\ u_v + \lambda_v] & \text{if} \quad u = u_v, \\ u + \lambda_v & \text{if} \quad u_v < u < u_p, \\ [u_p + \lambda_v\,,\ \infty] & \text{if} \quad u = u_p, \end{cases} \tag{6}$$

where u_v is the dimensionless evaporation temperature of the water, u_p is the dimensionless solid fuel pyrolysis temperature, and λ_v is the dimensionless evaporation heat related to the latent heat of water evaporation $\Lambda_v = 2.25 \times 10^6$ (J kg^{-1}) and the fuel moisture content M_v (kg of water/kg of dry fuel),

$$\lambda_v = \frac{M_v \Lambda_v}{C T_\infty} \tag{7}$$

It should be noted that the multivalued operator does not exactly represent the physical phenomena in the burnt zone because there is no longer any water vapour in the porous medium. This shortcoming is avoided by setting $\lambda_v = 0$ in the burnt area [10].

Equation (3) represents the loss of solid fuel due to combustion where

$$g(u) = \begin{cases} 0 \ if \ u < u_p, \\ \gamma \ if \ u = u_p, \end{cases} \tag{8}$$

that is, there is no loss of solid fuel if the temperature is below the pyrolysis temperature, and it remains constant when the temperature of pyrolysis is reached. This constant value is inversely proportional to the solid fuel half-life time $t_{1/2}$, of the combustion of each type of fuel,

$$\gamma = \frac{\ln 2[t]}{t_{1/2}}. \tag{9}$$

Finally, we now explain all the terms in the temperature equation, Eq. (1). First, the convective term $\beta \mathbf{v} \cdot \nabla e$ represents the energy convected by the gas pyrolysed through the elementary control volume. The surface wind velocity, $\mathbf{v}$, can either be considered as a constant based on data provided by the meteorological information available, or it can be computed by the wind model HDWind [5, 11, 12], developed by the authors and integrated accordingly in the same online GIS interface as the PhyFire model. The parameter β is a correction factor that is explained in [22]. In sum, as the PhyFire model is a single-phase model (mainly solid-phase), it can be considered a simplification of a gas-solid two-phase model in which the enthalpy transported in the gas phase is retained, and β represents the fraction of transported heat that must be taken into account in Eq. (1). For typical values of the heat capacity for the air and for the fuel (wood), β can be estimated to have an order of magnitude of 10^{-2} inside the flame, and 10^{-4} away from the flame.

Second, the zero-order term αu in Eq. (1), represents one of the 3D factors that the PhyFire model takes into account: the energy lost by natural convection in the vertical direction. Parameter α depends on the natural convection coefficient H

through the following expression:

$$\alpha = \frac{H[t]}{MC} \tag{10}$$

where $[t]$ is a time scale.

Third, the non-local radiation term r is another overriding 3D factor that represents the radiation from the flames above the surface where the fire takes place,

$$r = \frac{[t]}{MCT_{\infty}} R \tag{11}$$

where R means the incident energy at a point $\mathbf{x} = (x, y, h(x, y)) \in S$ due to the radiation coming from the flames above the surface, per unit of time and unit of area, reached by adding the contribution of all directions Ω on the hemisphere above the fuel layer,

$$R(\mathbf{x}) = \int_{\omega=0}^{2\pi} I(\mathbf{x}, \mathbf{\Omega})\mathbf{\Omega} \cdot \mathbf{N} \, d\omega, \tag{12}$$

I is the total radiation intensity, that is, the integral overall wavelength of the radiation energy passing through an area per unit of time, per unit of projected area, and per unit of solid angle, ω is the solid angle, and $\mathbf{N}$ is the unit normal vector to the surface S.

By omitting scattering following [25], the differential equation describing the total radiation intensity I at any position along a given path p in a grey medium may be written as

$$\frac{dI}{dp} + a(p)I(p) = a(p)I_b(p), \tag{13}$$

where I_b is the black body total radiation intensity and is governed by the Stefan-Boltzmann law, corresponding to the integral over all wavelengths of the emissive power of a black body

$$I_b = \frac{\sigma}{\pi} T^4, \tag{14}$$

where $\sigma = 5.6699 \times 10^{-8}\,\mathrm{w\,m^{-2}\,K^{-4}}$ is the Stefan-Boltzmann constant, and the temperature T reaches the flame temperature denoted by T_f. a is the radiation absorption coefficient inside the flame, and our model's third parameter.

Each incident energy contribution depends on the flame temperature T_f and flame length F, considering the flame tilt due to the surface slope and wind. In turn, experimental data from [3] show that the flame length clearly depends on wind speed and surface slope, and obviously on the fuel type. In [6], a global sensitivity analysis shows the advantage of introducing a flame length sub-model in terms of wind speed and surface slope,

$$F = (F_H + F_v|v|^{1/2})(1 + F_s s^2) \tag{15}$$

where F_H is an independent flame length parameter, F_v is a wind correction factor, F_s is a slope correction factor, $|v|$ is the wind speed, and s represents the slope at each point on the surface S. The first factor in Eq. (15) corresponds to the correction of flame length due to the wind, being based on the observation of the experimental curves for different fuels in [3], where the increase in the length of the flame due to the wind responds to such this function. F_H corresponds to zero wind, and the correction coefficient F_v has been added to experimentally adjust the different behaviours for each type of fuel by a least-squares method. The second factor in Eq. (15) corresponds to the correction of flame length according to surface slope, where again, a correction factor has again been added to this expression to adjust data from [3] in the least-squares sense. For further details see [6].

The main mechanisms of heat transfer in most wildfires are convection and radiation, which inform the fire spread by transmitting heat from the burning area to the unburned fuel in its path. This can be defined as a short-range fire propagation mechanism. In our model, convective heat transfer is depicted by the convective term $\beta \mathbf{v} \cdot \nabla e$ in Eq. (1), and radiative heat transfer by the non-local radiation term r on the right-hand side of Eq. (1). However, under certain conditions, fire spread can occur through mass transfer; in other words, by fire-spotting, which can be interpreted as a long-range fire propagation mechanism. The term q on the right-hand side of Eq. (1) stands for the random heat contribution due to fire-spotting. Section 2.3 details a way of representing the heat contribution due to fire-spotting is detailed, following the premises in [17, 21, 26].

2.2 Numerical Method

The numerical scheme used is based on P1 finite element approximation on a regular mesh, the semi-implicit Euler time scheme, and the Yosida approximation for the multivalued operator.

$\Delta t = t^{n+1} - t^n$ is a time step, let c^n, e^n and u^n denote approximations at time step t^n to the exact solution, c, e and u, respectively.

At each time step, we solve the following:

$$\frac{e^{n+1} - \bar{e}^n}{\Delta t} + \alpha u^{n+1} = r^n + q^n, \tag{16}$$

$$e^{n+1} \in G(u^{n+1}), \tag{17}$$

$$\frac{c^{n+1} - c^n}{\Delta t} = -g(u^{n+1})c^{n+1}., \tag{18}$$

where $\bar{e}^n = e^n \circ X^n$, and $X^n(\mathbf{x}) = X(\mathbf{x}, t^{n+1}, t^n) \approx \mathbf{x} - \beta \mathbf{v} \Delta t$ is the position at time t^n of the particle that is at position $\mathbf{x}$ at time t^{n+1}.

The basic premise is to treat the positive terms implicitly. The heat contribution of non-local radiation r and fire-spotting q heavily depends on the temperature u and on the fuel mass c, and it will therefore be explicitly evaluated at time t^n.

The multivalued operator in Eq. (17), is a maximal monotone that can be solved by the Yosida approximation (see [22] or [14]), obtaining

$$u^{n+1} = J_{1/\alpha \Delta t}\Big(\frac{1}{\alpha \Delta t}\bar{e}^n + \frac{1}{\alpha}(r^n + q^n)\Big) \quad (19)$$

Once u^{n+1} has been obtained by solving Eq. (19), we calculate e^{n+1} and c^{n+1} explicitly

$$e^{n+1} = \bar{e}^n - \alpha \Delta t u^{n+1} + \Delta t (r^n + q^n), \quad (20)$$

$$c^{n+1} = \frac{c^n}{1 + \Delta \tau g(u^{n+1})}. \quad (21)$$

Details of how to compute the heat received at each point due to radiation in each time step r^n can be found in [22] and [14], for both vertical and tilted flames. The calculation of the heat received due to fire-spotting is reported in Sect. 2.3.

To be a useful tool for supporting decision-making, any simulation model must provide results in a much shorter computation time than the simulated problem. To enhance the efficiency of the simulation process, different improvements have been implemented, with one of them being the definition of active nodes. We define a uniform and fine spatial mesh at the beginning of the numerical process, and for each time step Δt we define a set of active nodes formed by those located inside a sufficiently large surface of the fire front (burning area), and we solve the corresponding equations only in this set of nodes. This reduces the computation time, as we do not have to do any calculations when there is no change in the solution.

Note that Eqs. (19), (20), and (21) can be solved simultaneously in all the active nodes, and hence parallel computation can be used to shorten the computational time. Indeed, calculation of the variables u^{n+1}, e^{n+1} and c^{n+1} have been computed by means of a parallel calculation using the API OpenMP, that is, the loop over all active nodes is parallelized [1].

Attempts have also been made to reduce the operational cost by introducing a *fire-spotting index* N_q in order to significantly reduce the computation cost of the set of possible firebrand receiver nodes. In practice, this computation is done every $N_q \times \Delta t$, instead of every Δt.

2.3 Fire-Spotting Term

Fire-spotting refers to embers or firebrands that are carried by the wind and which start new fires (spot fires) beyond the main fire's direct ignition zone. These secondary fires frequently cause hazardous situations for firefighters and contribute to the increase in the rate of fire spread. The physical dynamics of fire behaviour, plume characteristics and atmospheric conditions around the fire are decisive factors affecting the generation of fire-spotting and the firebrands' landing patterns and flight paths. However, fire-spotting has a markedly random component, so existing fire-spotting models are necessarily probabilistic [19, 24, 26, 29].

Following a more thorough review of the research on fire-spotting, we propose a preliminary fire-spotting sub-model that closely fits the characteristics of the PhyFire model based on the ideas of the RandomFront 2.3 in [26], as a random heat contribution added to the right-hand side of Eq. (1). The heat contribution due to fire-spotting is written in terms of the distance of firebrand distribution $\phi(\ell)$

$$q = Q \times N_q \times \Delta_t \times \phi(\ell) \tag{22}$$

where following [24, 26], the firebrand landing distance ℓ can be assumed to follow a lognormal distribution

$$\phi(\ell) = \frac{1}{\ell\sigma\sqrt{2\pi}} \exp(-\frac{(\ln(\ell) - \mu)^2}{2\sigma^2}) \tag{23}$$

where μ and σ are the mean and standard deviation of the logarithm of the variable $\ln \ell$, that is, $\ln \ell \sim N(\mu, \sigma)$.

Q is a factor transforming the probability density function $\phi(\ell)$ into energy, N_q is a fire-spotting index introduced to reduce the computational cost of the fire-spotting module, and Δt is the time step of the time discretization used to numerically solve the PhyFire model Eqs. (1), (2), and (3), and is detailed in Sect. (2.2).

We computed the set of possible firebrand emitter nodes as a subset of the fire front, and then the set of possible firebrand receiver nodes from each emitter node so that the main direction of firebrand propagation is wind direction, and the firebrands' landing distance is computed in terms of the mean distance of firebrand landing, not more than 1.5 times this mean distance, and not to close to the fire-front.

Specifically, for each node m of the set of active nodes of the regular initial spatial mesh, the set S_E of possible emitting nodes m_E is selected as those nodes m that verify the following conditions on the dimensionless solid fuel temperature $u(m)$ at the node m, and the solid fuel mass fraction $c(m)$ at the node m,

$$u(m) \geq u_p = (T_p - T_\infty)/T_\infty \tag{24}$$

$$0.45 \leq c(m) \leq 0.55 \tag{25}$$

Then, the set $S_R(m_E)$ of possible firebrand receiver nodes from the emitter node m_E is then computed as a random subset of the set of nodes m_R verifying

$$\ell < 1.5\mu_l \tag{26}$$

$$\cos\alpha \geq 0.9 \tag{27}$$

$$\cos\alpha \geq \frac{\ell}{\sqrt{\ell^2 + dx^2}}, \tag{28}$$

where ℓ here represents the distance from the emitter node m_E and the receiver node m_R, dx is the spatial regular mesh size, and α is the angle formed by the segment $\overline{m_E m_R}$ and the wind direction at the emiser node m_E. Not all these firebrands trigger a new fire, so it is only considered a random fraction of the set described; that is, the effective firebrand is in fact a random subset of this set. In practice, a random number between 0 and 100 is generated for each possible receiving node; if the generated number is greater than $100\sqrt{10N_q}$, it is considered an effective receiver node, otherwise it is not.

From a computational point of view, looping through the nodes to locate the possible nodes receiving firebrands is very expensive, and as fire-spotting is random, we define an index N_q to compute the fire-spotting term each $N_q \times \Delta t$ instead of at each time step of time discretization.

3 Online GIS Interface

In order to automate the processes of input data capture and output data visualization during the simulation process, both PhyFire and HDWind were first integrated into a GIS [8] (Fig. 1). This GIS-based interface had a dual purpose: on the one hand, it provided a more accessible tool for a broader audience that might not be familiar with the model; and on the other hand, it facilitated the testing and validation process. The GIS tool chosen for this integration was ArcMap 10.4 of the Esri ArcGIS Desktop suite, and the interface was developed as a Python add-in for ArcMap. The functionality of each tool was implemented as a script using the Python programming language and the ArcPy geoprocessing library. They were developed for its use throughout Spain, so the scope of the spatial information currently used is limited to that country. For this purpose, a geodatabase has been developed containing the three maps needed for extracting the spatial information our models use: a first map containing the height of the surface, a second map gathering all the information related to fuel type, with both maps being used by the two models, and a third map collecting all the elements involving the function of either artificial or natural fire breaks that affect the fire spread, being used solely by the PhyFire model (see Fig. 2). The PhyFire integrated in the GIS tool uses the following input data: topography, fuel load and type, weather conditions, ignition location, and fire suppression tactics. It predicts the fire spread for the established

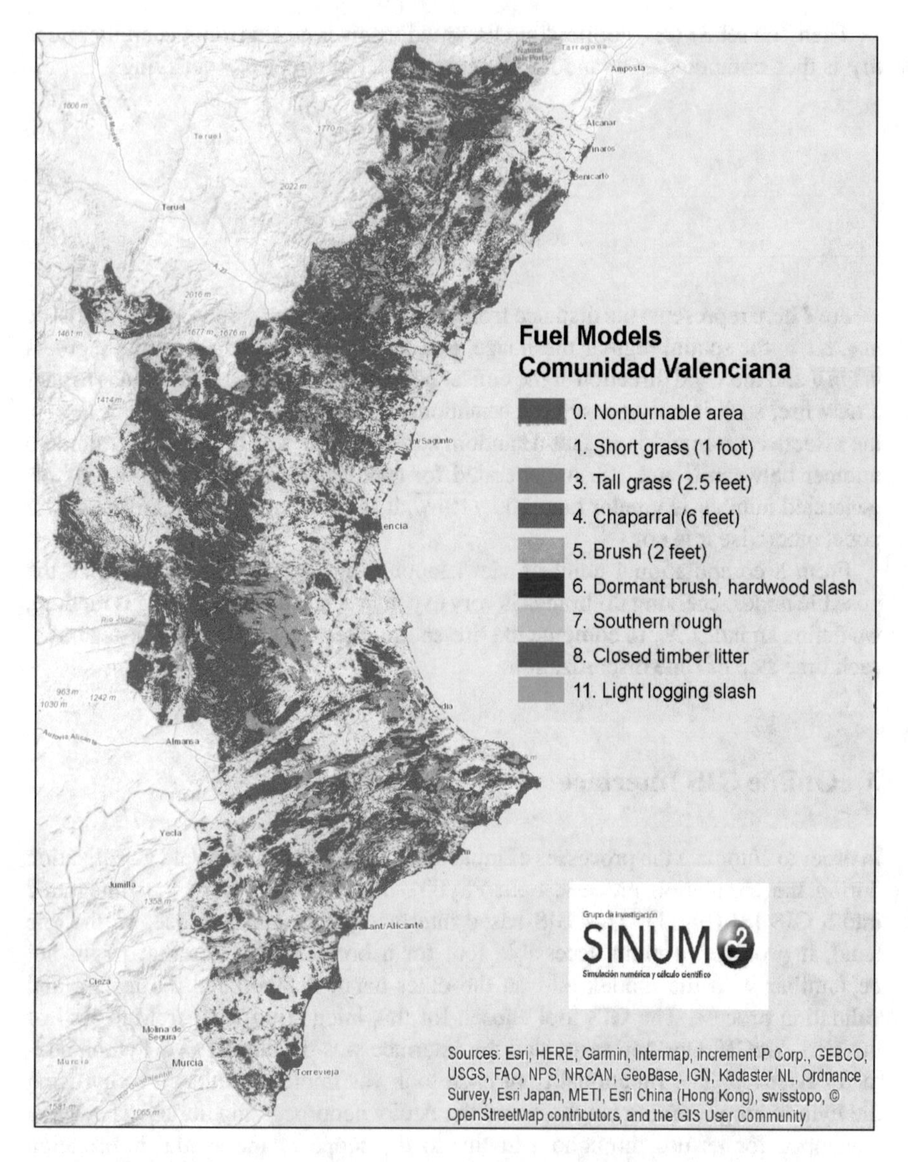

Fig. 1 Detail of the map of fuel types for the Valencian community. (Created using ArcGIS® software by Esri. (ArcGIS® and ArcMap™ are the intellectual property of Esri and are used herein under license. Copyright © Esri. All rights reserved. For more information about Esri® software, please visit www.esri.com))

time period, providing the following outputs at each time step: perimeter of the burnt area and position of the fire front.

The disadvantages of this integration in ArcGIS are that the Python add-in must be adapted to each ArcGIS version and the user must have the corresponding ArcGIS license. To overcome these disadvantages and make both PhyFire and

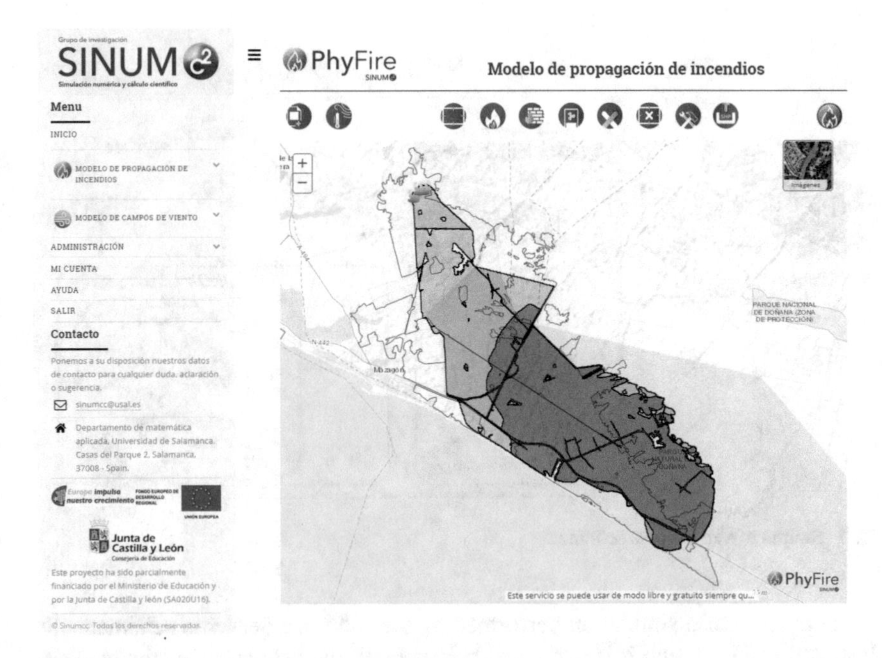

Fig. 2 Sinumcc Web Server displaying a simulation of the forest fire that affected the Doñana National Park in June 2017

HDWind, widely accessible tools, a web platform has been developed using current communication and data processing technologies, such as Api REST, JSON and ArcGIS Server. The platform http://sinumcc.usal.es allows uploading fire simulation data, and the pre-processing, processing and visualization of simulation results. The system carries out the phases of the process in a global way, providing the user with a quick visualization. The web platform involves two large modules: the Sinumcc Web Platform itself and the Sinumcc GIS Server. The Sinumcc Web Platform is used to collect data intuitively, semi-automatically and visually, and also to display the results of the simulations. The Sinumcc GIS Server is the module for performing these simulations based on the data collected from the Sinumcc Web Platform.

Data are exchanged between both modules by JavaScript Object Notation (JSON). Once the data provided by the user have been collected, they are stored in a database in JSON format. This not only allows storing the simulation's data and initial parameters, but also serves as a storage path for the Sinumcc GIS Server to collect, process and adapt these data so that the PhyFire model can perform the simulation. Once the simulation has been carried out and the results obtained, they are again saved in the database so that the Sinumcc Web Platform can display the results of the simulation. With this architecture, both modules interact asynchronously, allowing simulations to be carried out regardless of the time they take to run, without waiting from the Sinumcc Web Platform for the Sinumcc GIS Server to finish, whereby as many simulations can be performed simultaneously as the server supports.

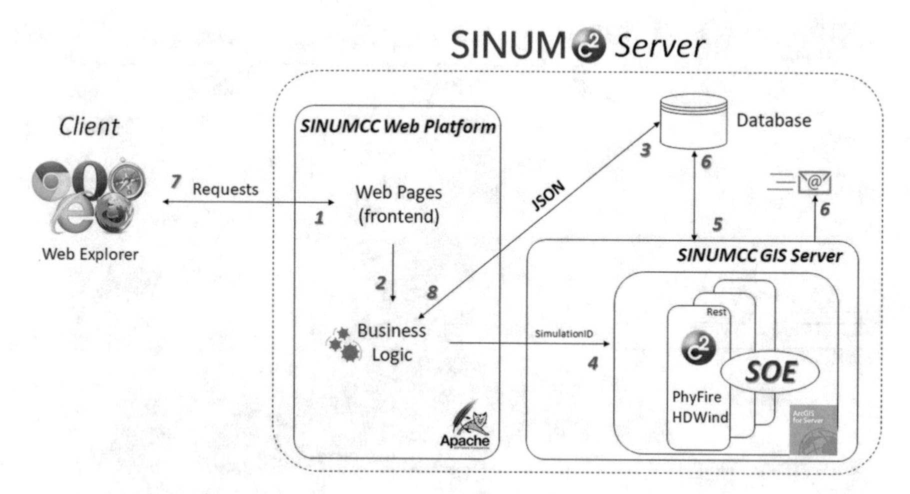

Fig. 3 Sinumcc Web process scheme

The process of a simulation performed by the Sinumcc Server is schematically illustrated in Fig. 3, and it is explained below. First, the user accesses the web site and loads the page with the content where the data about a simulation can be entered (see 1 in Fig. 3). When launching the simulation from the web platform, the business logic collects the data (see 2 in Fig. 3) and stores them in the database in JSON format (see 3 in Fig. 3). The simulation identification number (ID) where the data have been stored from the web platform is relayed to the Sinumcc GIS Server (see 4 in Fig. 3), which collects data from the database and performs the simulation using an instantiation of a Server Object Extension (SOE; see 5 in Fig. 3). As many instances can be created as requests made to the server. Once the simulation has been completed, the Sinumcc GIS Server stores the simulation results and notifies the user (see 6 in Fig. 3). Upon receipt of the notification that the simulation has ended, the user accesses the web platform and requests the simulation (see 7 in Fig. 3). Finally, the business logic collects the data in JSON format and displays them to the user (see 8 in Fig. 3).

3.1 Sinumcc Web Platform

The Sinumcc Web Platform follows a model-view-controller (MVC) design pattern. The model is the core component of this architectural arrangement; it manages the data, and receives user input from the controller independently of the user interface. The view is the particular display of information. The controller responds to the user's input and interacts with the data model through the business logic.

The following information is required for the Sinumcc Web Platform: Name, date and description of the fire to be simulated, selection of the simulation area limited to 25 km^2, basic weather information (temperature, humidity, wind), position of the initial source of the fire, and position of possible fire suppression tactics. Choosing the simulation area is the necessary prior step to obtain the topographic and distribution data and fuel type required by the PhyFire model from the available cartographic database. If only wind data (direction and intensity) are provided in the basic meteorological information dialog box, the wind is considered constant throughout the area. There is the possibility of incorporating geolocalized point wind data in the simulation zone, with the system calling upon the HDWind wind model to generate a local wind field that adjusts the point data and feeds the PhyFire model.

3.2 *Sinumcc GIS Server*

The Sinumcc GIS Server is responsible for performing the simulations themselves by developing an standard operating environment (SOE) that allows the simulations to be launched asynchronously, which means that as many simulations can be made as the server supports. The SOE consists of three modules: the management module, the SQL module, and the notification module; and it is developed in three stages: pre-process, process and post-process, as shown in Fig. 4.

The management module (see 1 in Fig. 4) is responsible for coordinating all of the modules involved in the SOE, defining the simulation's current status, and interacting with the SQL module and the notification module. The structured query language (SQL) module collects the input data (see 2 in Fig. 4) the PhyFire model

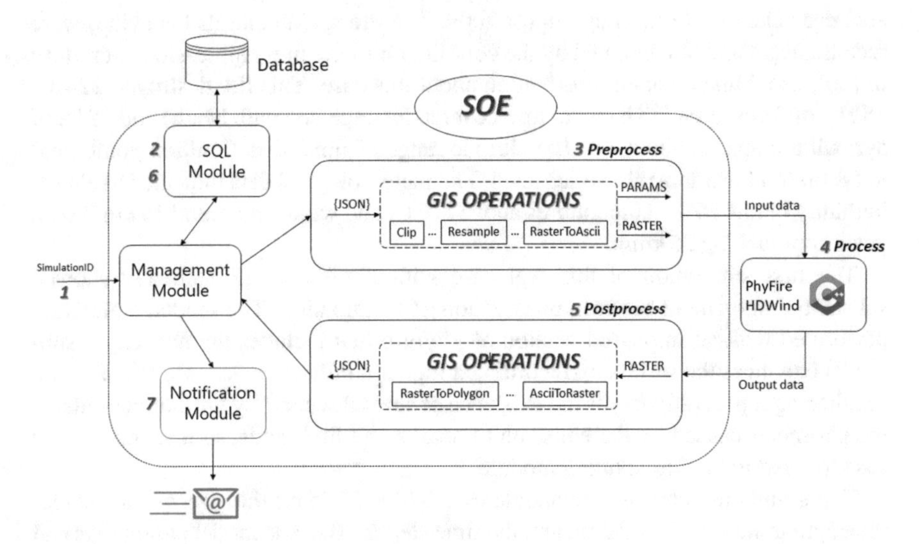

Fig. 4 Sinumcc GIS Server scheme

needs to perform the simulations and stores the provided results (see 6 in Fig. 4). The notification module (see 7 in Fig. 4) relates the user to the simulation launched, sending a notification when the simulation has finished or an error has occurred.

In the pre-processing stage (see 3 in Fig. 4), the data provided by the user in JSON format are transformed into the input files that the PhyFire model needs to perform the simulations, executing the corresponding GIS operations required, which are not detailed here as they exceed the aims of this paper. In addition to the input files corresponding to the PhyFire model parameters, four input files should be highlighted: the file corresponding to the topography of the simulation area; the file of the spatial distribution of fuel, which includes possible initial firewalls (initial conditions for the fuel), and the file of the spatial distribution of the types of fuels (these three files are obtained from the available geodatabase); and the fourth input file corresponds to the fire source or sources (initial temperature conditions). The process stage (see 4 in Fig. 4) corresponds to the running of the PhyFire model, and eventually of the HDWind model, with the input data provided. The PhyFire model generates two types of output files corresponding to the temperature and fuel that need to be transformed into JSON format during the post-process stage (see 5 in Fig. 4), so that the user can finally visualize the perimeter of the burnt area and position of the fire front for each time step.

4 Real Example

The new fire-spotting term is tested by simulating a real fire that we have already simulated with previous versions of the PhyFire model. This wildfire occurred in an area near Ourense (Spain) in August 2009. The fire spread and its behaviour were reconstructed and documented by the coordinator of the fire-suppression operations in [20], providing detailed information about this case. This fire destroyed 224 ha, 185 ha of forest area (83 ha were tree-covered interspersed with heath) and 39 ha of agricultural area in about 4 h. The altitude ranged from 540 m (ignition point area) to 680 m (end fire area) above sea level. The meteorological data indicated a relative humidity below 25%, temperatures above 30 °C, and winds of around 15 km/h with gusts approaching 25 km/h.

The first simulation of this real case with PhyFire in [23] already showed substantial agreement between observation and simulation. The second simulation performed with an improved version of PhyFire that includes the fire length sub-model enhances the results by recording a higher similarity index, see [6]. The fire monitoring report [20] details the existence of several secondary fire sources outside the perimeter caused by the emission of sparks and firebrands, so it seems a good case for testing the fire-spotting module.

The simulation area is a rectangle of $3{,}320 \times 2{,}745$ m, the size of the regular rectangular mesh is 7.5×7.5 m, and the time step is 30 s. The model parameters and input variables values are detailed in Tables 4, 5, and 6.

Table 4 Model parameters

Parameter	Symbol	Value
Natural convection coefficient	H	$15\,\mathrm{J\,s^{-1}\,m^{-2}\,K^{-1}}$
Convective term factor	β	0.015
Mean absorption coefficient	a	$0.095\,\mathrm{m^{-1}}$

Table 5 Fuel-type-dependent input variable values

Fuel type	M_0	M_v	T_f	T_p	$t_{1/2}$	C	F_H	F_v	F_s
Short grass	0.1	0%	1300	500	100	1800	0.2606	0.6001	5.4330
Timber grass	1.0	10%	1300	500	100	2000	1.1100	0.4712	0.6759
Brush	2.3	10%	1300	500	200	2300	3.7780	0.5075	2.8280
Dormant brush	2.2	10%	1300	500	200	2300	3.3240	0.4888	2.6880
Inflammable brush	2.4	15%	1300	500	300	2300	3.9320	0.6752	3.0150

Table 6 Fire-spotting module parameters

Parameter	Symbol	Value
Mean of logarithm of firebrand landing distance	μ	6.5
Standard deviation of logarithm of firebrand landing distance	σ	0.2
Fire-spotting index	N_q	10
PDF to energy factor	Q	11

Ambient temperature was around 30 °C and relative humidity around 28%. Wind data were provided at four points in the domain to adjust the wind field using the HDWind model, at a height of 10 m and updated every half hour. Wind intensity ranged from 2.54 to 4.75 m/s, and its direction shifted from east to northeast. The computing time on a laptop equipped with an Intel Core i7-2410M processor (two cores, each one working at a frequency of 1.8 GHz) and 4 GB RAM, was 12 min and 32 s for 4.5 h of simulation.

Figures 5 and 6 show the simulated perimeter after 1 h and 50 min and 2 h, respectively. This was the period of highest fire intensity in which fire-spotting caused new fires beyond the fire perimeter, as the simulation reflects.

5 Conclusions

This paper covers the most recent improvements made to the simplified physical fire spread model PhyFire developed by the research group on Numerical Simulation and Scientific Computing (SINUMCC) at the University of Salamanca. PhyFire now integrates a new module to simulate fire-spotting as a random heat contribution. The online GIS interface developed facilitates access to the model and the simulation process, automating the complex data input procedure and providing a graphical display of the results. The simulation of a real fire provides a good case study of the above.

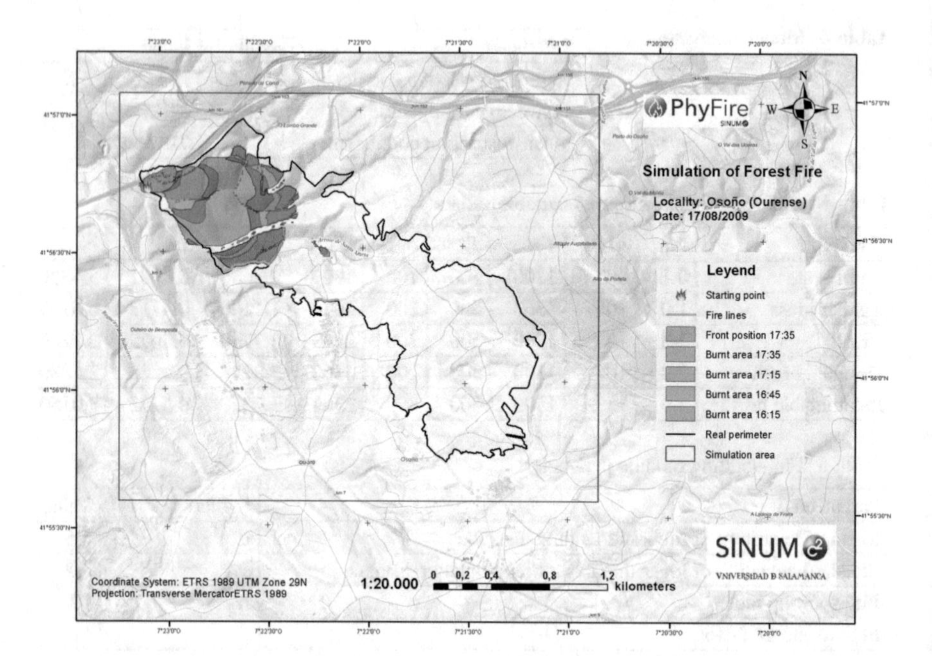

Fig. 5 Osoño: simulated burnt area (grey) and active front (orange) after 1 h and 50 min. A new fire has started due to fire-spotting. The black line is the real final perimeter

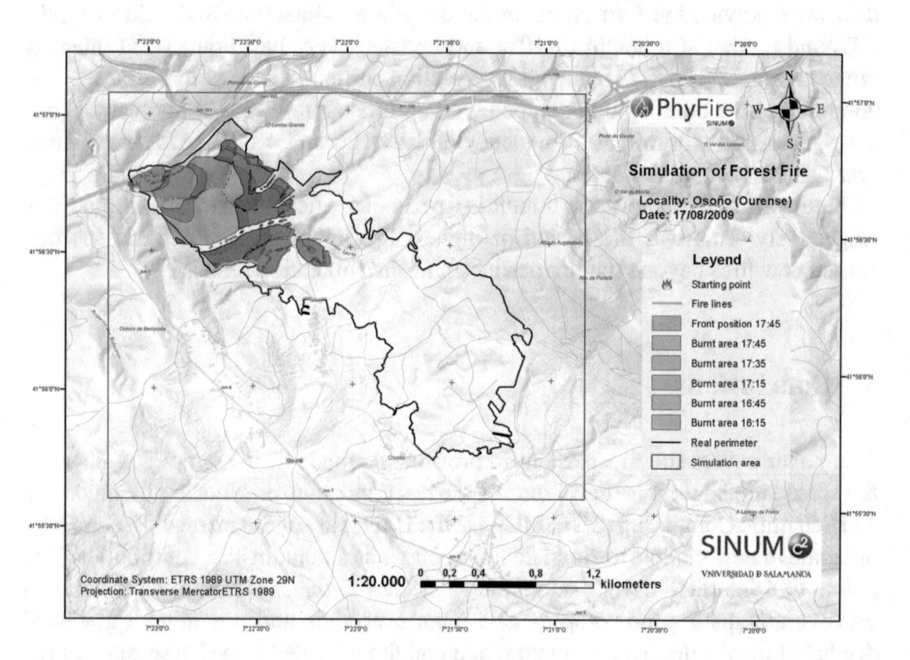

Fig. 6 Osoño: simulated burnt area (grey) and active front (orange) after 2 h, showing the evolution of the secondary fire due to fire-spotting

Acknowledgments This work has been partially supported by the Conserjería de Educación (Department of Education) of the regional government, the *Junta de Castilla y León* (SA020U16), by the University of Salamanca General Foundation (TCUE Grant and Prototransfer) both with the participation of ERDF, and by *Fundación Universidades y Enseñanzas Superiores de Castilla y León* through the University Nursery Business Promoters' first award in 2018.

G. Pagnini is supported by the Basque Government through the BERC 2018–2021 program and by Spanish Ministry of Economy and Competitiveness MINECO through BCAM Severo Ochoa excellence accreditation SEV-2017-0718 and through project MTM16-76016-R MIP.

We thank the ICIAM2019 organization for the opportunity to disclose the modelling of environmental issues and highlight the role of Applied Mathematics in improving the environment through the mini-symposium specifically dedicated to environmental problems.

References

1. Álvarez, D., Prieto, D., Asensio, M.I., Cascón, J.M., Ferragut, L.: Parallel implementation of a simplified semi-physical wildland fire spread model using openMP. In: Martínez de Pisón, F., Urraca, R., Quintián, H., Corchado, E. (eds.) Hybrid Artificial Intelligent Systems. HAIS 2017. Lecture Notes in Computer Science, vol. 10334. Springer, Cham (2017). https://doi.org/10.1007/978-3-319-59650-1_22
2. Aponte, C., de Groot, W., Wotton, B.: Forest fires and climate change: causes, consequences and management option. Int. J. Wildland Fire **25**(8), I–II (2016). https://doi.org/10.1071/WFv25n8_FO
3. Arellano, S., Vega, J., Ruíz, A., Arellano, A., Álvarez, J., Vega, D., Pérez, E.: Foto-guía de combustibles forestales de Galicia. Versión I, Andavira Editora, S.L. (2016). https://doi.org/10.14195/978-989-26-16-506
4. Asensio, M.I., Ferragut, L.: On a wildland fire model with radiation. Int. J. Numer. Methods Eng. **54**(1), 137–157 (2002). https://doi.org/10.1002/nme.420
5. Asensio, M.I., Ferragut, L., Simon, J.: A convection model for fire spread simulation. Appl. Math. Lett. **18**, 673–677 (2005). https://doi.org/10.1016/j.aml.2004.04.011
6. Asensio, M.I., Santos-Martín, M.T., Álvarez-León, D., Ferragut, L.: Global sensitivity analysis of fuel-type-dependent input variables of a simplified physical fire spread model. Math. Comput. Simul. **172**, 33–44 (2020). https://doi.org/10.1016/j.matcom.2020.01.001
7. Cascón, J.M., Ferragut, L., Asensio, M.I., Prieto, D., Álvarez, D.: Neptuno ++: an adaptive finite element toolbox for numerical simulation of environmental problems. In: XVIII Spanish-French School Jacques- Louis Lions about Numerical Simulation in Physics and Engineering, Las Palmas de Gran Canaria (2018). http://hdl.handle.net/10366/138180
8. Ferragut, L., Asensio, M.I., Montenegro, R., Plaza, A., Winter, G., Serón, F.J.: A model for fire simulation in landscapes. In: Désidéri y otros, J. A. (eds.) Third ECCOMAS Computational Fluid Dynamics Conference, París (Francia), Sept 1996, pp. 111–116. John Wiley & Sons
9. Ferragut, L., Asensio, M.I., Monedero, S.: Modelling radiation and moisture content in fire spread. Commun. Numer. Methods Eng. **23**(9), 819–833 (2007). https://doi.org/10.1002/cnm.927
10. Ferragut, L., Asensio, M.I., Monedero, S.: A numerical method for solving convection-reaction-diffusion multivalued equations in fire spread modelling. Adv. Eng. Softw. **38**(6), 366–371 (2007). https://doi.org/10.1016/j.advengsoft.2006.09.007
11. Ferragut, L., Montenegro, R., Montero, G., Rodríguez, E., Asensio, M.I., Escobar, J.: Comparison between 2.5-D and 3-D realistic models for wind field adjustment. J. Wind Eng. Indus. Aerodyn. **98**, 548–558 (2010). https://doi.org/10.1016/j.jweia.2010.04.004
12. Ferragut, L., Asensio, M.I., Simon, J.: High definition local adjustment model of 3D wind fields performing only 2D computations. Int. J. Numer. Methods Biomed. Eng. **27**, 510–523 (2011). https://doi.org/10.1002/cnm.1314

13. Ferragut, L., Asensio, M.I., Cascón, J.M., Prieto, D.: A simplified wildland fire model applied to a real case. In: Advances in Differential Equations and Applications. SEMA SIMAI Springer Series, vol 4, pp. 155–167. Springer International Publishing, Cham (2014). https://doi.org/10.1007/978-3-319-06953-1_16
14. Ferragut, L., Asensio, M.I., Cascón, J.M., Prieto, D.: A wildland fire physical model well suited to data assimilation. Pure Appl. Geophys. **172**(1), 121–139 (2015). https://doi.org/10.1007/s00024-014-0893-9
15. Finney, M.: FARSITE: fire area simulator-model development and evaluation. Research Paper RMRS-RP-4 (revised), U.S. Department of Agriculture, Forest Service, Rocky Mountain Research Station, Ogden (2004)
16. Jolly, W.M., Cochrane, M.A., Freeborn, P.H., Holden, Z.A., Brown, T.J., Williamson, G.J., Bowman, D.M.J.S.: Climate-induced variations in global wildfire danger from 1979 to 2013. Nat. Commun. **6**(7537) (2015). https://doi.org/10.1038/ncomms8537
17. Kaur I., Mentrelli A., Bosseur F., Filippi J.B., Pagnini G.: Turbulence and fire-spotting effects into wildland fire simulators. Commun. Nonlinear Sci. Numer. Simul. **39**, 300–320 (2016). https://doi.org/10.1016/j.cnsns.2016.03.003
18. Mandel, J., Beezley, J.D., Kochanski, A.K.: Coupled atmosphere wildland fire modeling with WRF 3.3 and SFIRE 2011. Geosci. Model Develop. **4**(3), 591–610 (2011). https://doi.org/10.5194/gmd-4-591-2011
19. Martin, J., Hillen, T.: The spotting distribution of wildfires. Appl. Sci. **6**, 177–210 (2016). https://doi.org/10.3390/app6060177
20. Morillo, A., Análisis del comportamiento del fuego forestal observado y simulado: estudio del caso del incendio forestal de Osoño (Vilardevós)-Verín-Ourense. Master of Advanced Studies dissertation, Higher Polytechnical College of Lugo, University of Santiago de Compostela (2011) (in Spanish)
21. Pagnini, G., Mentrelli, A.: Modelling wildland fire propagation by tracking random fronts. Nat. Hazards Earth Syst. Sci. **14**, 2249–2263 (2014). https://doi.org/10.5194/nhess-14-2249-2014
22. Prieto, D., Asensio, M.I., Ferragut, L., Cascón, J.M.: Sensitivity analysis and parameter adjustment in a simplified physical wildland fire model. Adv. Eng. Softw. **90**, 98–106 (2015). https://doi.org/10.1016/j.advengsoft.2015.08.001
23. Prieto, D., Asensio, M.I., Ferragut, L., Cascón, J.M., Morillo, A.: A GIS based fire spread simulator integrating a simplified physical wildland fire model and a wind field model. Int. J. Geograph. Inf. Sci. **31**(11), 2142–2163 (2017). https://doi.org/10.1080/13658816.2017.1334889
24. Sardoy, N., Consalvi, J.L., Kaiss, A., Fernandez-Pello, A.C., Porterie, B.: Numerical study of ground-level distribution of firebrands generated by line fire. Combust. Flame **154**, 478–488 (2008). https://doi.org/10.1016/j.combustflame.2008.05.006
25. Siegel, R., Howell, J.R.: Thermal Radiation Heat Transfer. McGraw-Hill Inc., New York (1972)
26. Trucchia, A., Egorova, V., Butenko, A., Kaur, I., Pagnini, G.: RandomFront 2.3 a physical parametrization of fire-spotting for operational fire spread models: implementation in WRF-Sfire and response analysis with LSFire+. Geosci. Model Develop. **12**(1), 69–87 (2019). https://doi.org/10.5194/gmd-12-69-2019
27. Tymstra, C., Bryce, R., Wotton, B., Taylor, S., Armitage, O.: Development and structure of Prometheus: the Canadian wildland fire growth simulation model, Information Report NOR-X-417, Canadian Forest Service, Northern Forestry Centre (2010). https://d1ied5g1xfgpx8.cloudfront.net/pdfs/31775.pdf
28. Vasconcelos, M., Guertin, D.: Firemap – simulation of fire growth with a geographic information system. Int. J. Wildland Fire **2**(2), 87–96 (1992). https://doi.org/10.1071/WF9920087
29. Wang, H.H.: Analysis on downwind distribution of firebrands sourced from a wildland fire. Fire Technol. **47**, 321–340 (2011). https://doi.org/10.1007/s10694-009-0134-4

Physical Parametrisation of Fire-Spotting for Operational Wildfire Simulators

Vera N. Egorova, Andrea Trucchia, and Gianni Pagnini

Abstract Fire-spotting is strongly affected by mean wind and fire intensity, not only because they characterise the transport of firebrands, but, also, because they change the geometry of the flame, namely, the flame height and the flame length. Interdependencies between the flame length and the fire intensity are discussed in literature by a number of empirical relations. In the present study, the energy conservation principle and the energy flow rate in the convection column above the fire line are considered in order to establish the relation between the flame geometry and the fire line intensity in wildfires. Moreover, in opposition to literature, the derived formula allows for stating the rate of spread of the fire propagation in terms of the flame geometry factors by taking into account also the effects of the horizontal mean wind and the terrain slope. Numerical examples show that fire-spotting is strongly impacted by the flame geometry, which is specified by the fuel and vegetation, and then it cannot be neglected in the physical parametrisation of the phenomenon.

V. N. Egorova
Departamento de Matemática Aplicada y Ciencias de la Computación, Universidad de Cantabria, Santander, Spain
e-mail: vera.egorova@unican.es

A. Trucchia
CIMA Research Foundation, Savona, Italia
e-mail: andrea.trucchia@cimafoundation.org

G. Pagnini (✉)
BCAM–Basque Center for Applied Mathematics, Bilbao, Spain
Ikerbasque–Basque Foundation for Science, Bilbao, Spain
e-mail: gpagnini@bcamath.org

M. I. Asensio et al. (eds.), *Applied Mathematics for Environmental Problems*, SEMA SIMAI Springer Series 6, https://doi.org/10.1007/978-3-030-61795-0_2

1 Introduction

Fire-spotting as a part of wildfire behaviour is a challenging multiscale physical problem [13]. Fire-spotting causes the acceleration of the rate of spread (ROS) [19] and therefore it is crucial in modelling fire propagation and it cannot be disregarded.

Due to its unpredictable nature, fire-spotting is here considered with a statistical approach. Following the approach already proposed by this research group (see, e.g., [12, 17, 24, 33]), fire-spotting can be included into an existing fire propagation model as a post-processing routine by a proper probability density function (PDF). Martin & Hillen [19] studied in detail the spotting distribution by taking into account launching and landing distributions. Kaur & Pagnini [16] proposed a physical parametrisation of the fire-spotting by taking into account maximum loftable height of the firebrand, mean wind and fire intensity. Wang [37] studied the downwind distribution of firebrands by considering the maximum travel distance that depends also on the geometrical characteristics of the flame.

Literature results show that the spreading of the fire is strongly affected by the geometrical characteristics of the flame [36]. Accurate estimation of the corresponding parameters allow to determine how a wildfire may be controlled: in fact the flame length is used to determine the size of fire control lines [23] and the flame height is used to predict the heat flux exposure [25]. Fire-spotting is also impacted by the fire intensity, which is a fundamental descriptor of wildfires and it is used by practitioners to predict probability of house survival during bush fires [40].

Establishing indicators for the onset of erratic or unexpected wildfire behaviour is an important endeavour, and flame characteristics are fundamental features for the determination of the combustion regimes [21]. Moreover, flame geometry is a descriptor of the surrounding vegetation and for this reason it is taken into account in fire-fighting strategies [9]. Since Byram's formula [7] a number of empirical formulae for the interdependency between flame length and fire intensity has been proposed [3]. The relationships between flame geometry and Froude number, or convection number, are studied since Nelson Jr.'s paper [20] in many recent works (see [30, 31]) together with the experimental approbation [38, 39].

But, in this respect, there is also an important lack in the literature on the relation between the flame height and the fire line intensity. Usually, this relation is established empirically by using statistical methods for a concrete case, and there is only a few attempts to develop a physical model. The first one is done by Albini in 1981 [1], that was improved by Nelson Jr. and co-authors in 2012 by including the entrainment [21]. The Albini's formula provides an estimation of the flame height by relating the flame height with the fire line intensity in the steady or light wind case. The flame height derived in this model emerges to be proportional to the fire line intensity, while the flame length results to be proportional to the fire line intensity to the power $2/3$, that means that the flame tilt results to be dependent on the fire line intensity. Later, this result has been extended by Nelson Jr. and co-authors [21] by formulating flame characteristics equations and by taking into account the entrainment velocity for low-wind fires [21]. The different power-law dependence between the two formulae follows from the inclusion of the characteristic buoyant velocity, that is dependent on the fire line intensity with the power $1/3$.

A further approach was proposed by Marcelli and co-authors on the basis of the radiative flux and the flame height is defined as the height of the equivalent radiant panel [18]. Another model based on radiation, that takes into account the moisture content and energy losses, was also proposed by Ferragut and collaborators [14].

Hence, motivated by this important lack, in the following we establish theoretically a formula for estimating the flame height and the flame length in wildfires by relating it with the fire line intensity in steady and unsteady cases. The derivation is based on the energy conservation principle and the concept of the energy flow rate in the convection column above a fire line, this last concept was originally introduced by Byram in 1959 [8]. In order to take into account also the impact of the wind and the slope on the flame geometry, the proposed formula is incorporated into Rothermel's ROS model [5, 26] and this impact emerges to be described by the same correcting factor, showing that when they augment the ROS the flame length enlarges.

Afterwards, we include the flame geometry into the firebrand landing distribution following the approach described in Reference [12]. Within the proposed parametrisation, numerical simulations show that the flame geometry, and in particular the flame length, contributes significantly to the generation of independent secondary fires.

The rest of the chapter is organised as follows. Section 2 deals with the derivation of the flame length-fire intensity relation on the basis of the energy conservation principle. The inclusion of this relation into the fire-spotting model is discussed in Sect. 3. In Sect. 4 the results of the numerical simulations are reported, and the final remarks are given in Sect. 5.

2 Flame Geometry and Fire Intensity Interdependence

In the present derivation, we assume that the main flame characteristics are the flame length L_{f}, the flame height h and the flame tilt θ, that are connected by the following trigonometric relation: $h = L_{\mathrm{f}} \cos\theta$.

It is natural that the flame height is strongly affected by the wind and the fire intensity. In the proposed formulation of the flame height, we assume that the impact of the fire intensity is represented mainly by the flame length, while effects of wind are capable to change the whole flame geometry in the sense that increasing wind speed provokes the increasing flame length and flame angle, resulting in decreasing flame height.

To the best of our knowledge, the first model in literature for the wind-blown turbulent flame from a line fire is the one proposed by Albini in 1981 [1], that leads to the following formula of flame height h:

$$h = \frac{(m_t - m_g)}{\rho \eta U} = \frac{T_g - T_t}{T_t - T_a} \frac{m_g}{\rho \eta U}, \tag{1}$$

where m and T are the mass flow of the flame fluid and the temperature, respectively, at a generic height z such that m_t and T_t denote the corresponding measurements at the flame top, that is, $z = h$, and m_g and T_g at the ground level, that is, $z = 0$, moreover T_a is the ambient air temperature, ρ is the air density, U is the wind speed, and η is the fraction of impinging air stream incorporated into flame fluid flow.

In 2012, Nelson Jr. and co-authors [21] followed the same formulation by Albini but they introduced the fire line intensity I_f through the following formula:

$$I_f = m_g c_p (T_g - T_a), \tag{2}$$

where c_p is the specific heat of air at constant pressure, and finally the flame height takes the following form:

$$h = \frac{T_a(T_g - T_t)}{\alpha\,(T_t - T_a)(T_g - T_a)} \left[\frac{1}{2g(\rho c_p T_a)^2}\right]^{1/3} I_f^{2/3}, \tag{3}$$

where α is the entrainment constant. The factor $(T_g - T_t)/(T_t - T_a)$ in (3) is stated equal to 1 by Nelson Jr. and co-authors [21], such that $T_t = (T_g + T_a)/2$ which is not true in general.

Derivation of (3) by Nelson Jr. and co-authors [21] is based on an ad hoc assumption that connects the horizontal wind speed and the (vertical) characteristic buoyant velocity which are indeed two independent quantities, and this independence is reflected by the necessity to use other two independent parameters, namely η and α. Thus, we propose a relation between the flame height and the fire line intensity in wildfires based on the energy conservation principle and the energy flow rate in the convection column above the fire line.

2.1 *The Energy Conservation Principle for the Estimation of the Flame Height*

In order to determine the flame height h, we consider an air parcel located at the top of the flame $z = h$, that is initially not buoyant, that is, the vertical velocity w is equal to 0, and it is heated by the flame. From the conservation of energy, we have

$$e + PV + H - [e_0 + P_0 V_0 + H_0] = Q - W_{sh}, \tag{4}$$

where e is the internal energy of the gas, P and V are the pressure and the volume, H is the mechanical energy, Q is the heat transferred into the gas and W_{sh} is the shaft work used to move the fluid. Terms with subscript 0 refer to the initial instant and those without it to a generic instant. The initial mechanical energy is

$$H_0 = gh \,, \tag{5}$$

that turns into

$$H = g\,(h + \delta h) + \frac{w^2}{2} \,, \tag{6}$$

where g is the acceleration of gravity and δh is the vertical displacement done by the air parcel. The work done on the gas W is stated equal to the work necessary to balance the gravity force, that is,

$$W = PV - P_0 V_0 + W_{sh} = -g\,(h + \delta h) \,, \tag{7}$$

and the heat transferred into the gas is stated equal to the increase of the internal energy, that is,

$$Q = e - e_0 \,. \tag{8}$$

Plugging all the above formulae into (4) we have that the vertical velocity due to the convection above the fire line is

$$|w| = \sqrt{2gh} \,. \tag{9}$$

Conversion of turbulent kinetic energy into heat, namely the turbulent kinetic energy dissipation ε, may also be included as a sink in (6), that is, $H \to H - \varepsilon$, and as a source in (8), that is, $Q \to Q + \varepsilon$, and formula (9) is still obtained.

In order to estimate the vertical velocity $|w|$, we consider now the energy flow rate in the convection column above a line of fire, hereinafter denoted by P_f, that is defined as the rate at which thermal energy is converted into kinetic energy in the convection column at a specified height z [8, 20]. In formulae, we have

$$P_f(z) = \frac{g I_f}{c_p T_a} = \frac{1}{2}\,\rho\, w^2 |w| = \frac{1}{2}\,\rho\, |w|^3 \,. \tag{10}$$

Finally, by plugging (9) into (10) we have the following estimation of the flame height in steady fires. In that case, flame height with no wind is assumed to be equal to the flame length due to the weak influence of wind on the flame length [22, 32]. Thus,

$$L_{f_0} = h_0 = \left[\frac{1}{2g(\rho c_p T_a)^2}\right]^{1/3} I_{f_0}^{2/3} \,, \tag{11}$$

where subscript 0 stands for the absence of wind. Formula (11) straightforwardly follows from the application of the energy conservation principle and the concept of energy flow rate in the convection column above a fire line.

From the trigonometric relation of the flame characteristics, in the presence of wind one gets

$$h = L_\mathrm{f} \cos\theta = \cos\theta \left[\frac{1}{2g(\rho c_\mathrm{p} T_\mathrm{a})^2} \right]^{1/3} I_\mathrm{f}^{2/3} . \tag{12}$$

From formula (11), h depends on fire intensity with the power $2/3$ in agreement with experimental measurements in literature [3] and from formula (3) it follows that

$$\cos\theta = \frac{T_\mathrm{a}(T_\mathrm{g} - T_\mathrm{t})}{\alpha\,(T_\mathrm{t} - T_\mathrm{a})(T_\mathrm{g} - T_\mathrm{a})} . \tag{13}$$

Within this framework, the vertical velocity of the air parcel results to be constant. This means that (10) refers to a vertical interval where the value of the product ρT_a is almost constant. Since, from the ideal gas law, we have that $\rho c_\mathrm{p} T_\mathrm{a} = c_\mathrm{p} P/R_\mathrm{a}$, where R_a is the gas constant per unit mass of air, then the pressure is almost constant. Hence, if we consider, for example, the hydrostatic balance, that is, $P = -\rho g z$, the present formalism holds when the approximation $\rho \sim 1/z$ holds.

2.2 *Entrainment Estimation*

In general, by using (10), formula (12) can be re-written as

$$h = \cos\theta \left[\frac{1}{2g^3\rho^2} \right]^{1/3} P_\mathrm{f}^{2/3} , \tag{14}$$

where the horizontal energy flow also affects the flame height through the flame angle θ.

Byram introduced the concept of energy flow rate in the convection column above a fire line and also that of energy flow rate in the wind field [8, 20]. The energy flow rate in the wind field, hereinafter denoted by P_w, is the rate of flow of kinetic energy through a vertical plane of unit area in a neutrally stable atmosphere at the height z specified for P_f, that is,

$$P_\mathrm{w}(z) = \frac{1}{2}\rho\,(U - V_\mathrm{ROS})^2\,|U - V_\mathrm{ROS}| = \frac{1}{2}\rho\,|U - V_\mathrm{ROS}|^3 , \tag{15}$$

where V_ROS is the ROS. Byram proposed to use the ratio $\kappa = P_\mathrm{f}/P_\mathrm{w}$ to characterise wildfires such that this ratio is also called the Byram's energy criterion [8]. Byram pointed out that this ratio can be useful in understanding and predicting the onset of erratic fire behaviour and the occurrence of blowup fires. In particular, a strong relationship has been observed between the occurrence of blowup fires and values of this ratio close to 1 [29]. When this ratio is close to 1, horizontal and vertical forcing are balanced and then the propagation is not mainly driven by one or by the other forcing. In this situation, fluctuations govern the motion and an erratic behaviour follows. Ratio κ can be related with the so-called convective Froude number [29].

Let us consider Byram's energy criterion, then from (10) and (15) we have the following equalities:

$$\kappa = \frac{|w|^3}{|U - V_{\mathrm{ROS}}|^3} = \frac{2 g I_{\mathrm{f}}}{\rho c_{\mathrm{p}} T_{\mathrm{a}} \, |U - V_{\mathrm{ROS}}|^3} \,. \tag{16}$$

From left side of formula (16), it holds

$$|w| = \kappa^{1/3} \, |U - V_{\mathrm{ROS}}| \,. \tag{17}$$

The entrainment can be roughly understood as the mixing between the ambient air and the rising plume of hot air above the fire line. From this point of view, the ratio between the horizontal mean wind U and the quantity $|U - V_{\mathrm{ROS}}|$ states how much the horizontal mean flow enters into the rising column of the fire-heated hot air. Hence, by remembering that η is the fraction of impinging air stream incorporated into flame fluid flow, that is,

$$\eta U = |U - V_{\mathrm{ROS}}| \,, \tag{18}$$

by using formula (18) we have a number of results related with the entrainment.

Combining (17) and (10), from (18) we obtain

$$\eta U = |U - V_{\mathrm{ROS}}| = \frac{|w|}{\kappa^{1/3}} = \frac{1}{\kappa^{1/3}} \left(\frac{2 g I_{\mathrm{f}}}{\rho c_{\mathrm{p}} T_{\mathrm{a}}} \right)^{1/3} \,, \tag{19}$$

that, compared against formula (20) in Reference [21], gives $\alpha = \kappa^{-1/3}$, where α is the entrainment constant by Nelson Jr. and co-authors [21], and finally it holds

$$h = \kappa^{1/3} \frac{T_{\mathrm{a}}(T_{\mathrm{g}} - T_{\mathrm{t}})}{(T_{\mathrm{t}} - T_{\mathrm{a}})(T_{\mathrm{g}} - T_{\mathrm{a}})} \left[\frac{1}{2g(\rho c_{\mathrm{p}} T_{\mathrm{a}})^2} \right]^{1/3} I_{\mathrm{f}}^{2/3} \,. \tag{20}$$

Comparing formulae (3), (12) and (20) we have the novel result

$$\cos\theta = \frac{T_{\mathrm{a}}(T_{\mathrm{g}} - T_{\mathrm{t}})}{(T_{\mathrm{t}} - T_{\mathrm{a}})(T_{\mathrm{g}} - T_{\mathrm{a}})} \kappa^{1/3} \,, \tag{21}$$

that establishes a new method to measure the flame angle θ, which is a local geometric information, in terms of the temperature and Byram ratio, which are general characteristics of the fire, such that an *effective* estimation is indeed provided. This is a positive property of formula (21) in view of its application in operational simulators of wildfires. From formula (21), it emerges that the tilting angle is independent of the fire intensity and its variability is mainly due to the wind and to the slope. Thus, this factor embeds both a macro-scale feature due to the wind and a meso-scale feature due to the terrain slope. In fact, by replacing κ in (21) with the left side of formula (16), it follows that the increasing of the wind reduces $\cos\theta$. Moreover, since $\cos\theta = 1$ can be achieved only with a flat terrain, formula (21) provides also a no-slope condition. This means that an explicit formula of $\cos\theta$ is crucial in modelling unsteady and realistic fires and, in general, it cannot be approximated to unity.

There are also some other results related with those previously derived. In particular, we observe that when formula (17) is plugged into the right side of (16) gives again (12). This derivation states as a further result that the dependence on z of the ratio κ is fully driven by the dependence on z of the quantity $|U - V_{\text{ROS}}|^{-3}$.

Moreover, if the total kinetic energy is considered, that is, $K = P_{\text{f}}/|w| + P_{\text{w}}/\,|U - V_{\text{ROS}}|$, then it holds

$$h = \frac{1}{\rho g}\frac{P_{\text{f}}}{|w|} = \frac{1}{g}\left[\frac{K}{\rho} - \frac{(U - V_{\text{ROS}})^2}{2}\right], \tag{22}$$

that shows how the flame height decreases when the wind increases.

2.3 Flame Geometry in the Unsteady Case

Formula (11) for the flame length holds only in the steady case. In order to generalise (11) by including the effects of the wind, and also of the slope, we consider the linear relation between the ROS and the fire line intensity as established by the Byram's formula [2, 7]. Later we recast the fire line intensity by the ROS from the Rothermel's model [5, 26], where an increasing factor due to wind and slope is employed.

Rothermel's model [5, 26] is the most widely used model in fire management systems and wildfire theory. It is a surface fire spread model based on the heat balance and the ROS is computed by

$$V_{\text{ROS}} = V_{\text{ROS}_0}(1 + \phi_{\text{wind}} + \phi_{\text{slope}})\,, \tag{23}$$

where coefficients ϕ_{wind} and ϕ_{slope} refer to wind and slope effects, whose values can be get, for example, from Reference [5]. If I_{f_0} is the fire intensity in the steady case, the classical Byram's formula reads [2, 7] $I_{\text{f}_0} = \varphi R V_{\text{ROS}_0}$, where φ is the net low heat of combustion and R is the fuel consumed in the active flaming front, and then for the unsteady case one gets

$$I_{\rm f} = \rho c_{\rm p} T_{\rm a} \sqrt{2g}\, L_{\rm f_0}^{3/2} (1 + \phi_{\rm wind} + \phi_{\rm slope}) \,, \tag{24}$$

where the proposed relation between the flame length and the fire intensity (11) is used, and $\phi_{\rm wind}$ and $\phi_{\rm slope}$ are exactly those adopted for the ROS.

Hence, the effects of wind and slope to the flame length can be included as follows:

$$L_{\rm f} = L_{\rm f_0} (1 + \phi_{\rm wind} + \phi_{\rm slope})^{2/3} \,. \tag{25}$$

For the steady fire on flat terrain under very light wind conditions Byram's formula holds, as well as equality (11). Thus,

$$I_{\rm f_0} = \varphi R V_{\rm ROS_0} = \rho c_{\rm p} T_{\rm a} \sqrt{2g}\, L_{\rm f_0}^{3/2} \,, \tag{26}$$

which leads to the following:

$$V_{\rm ROS_0} = \frac{\rho c_{\rm p} T_{\rm a}}{\varphi R} \sqrt{2g}\, L_{\rm f_0}^{3/2} \,. \tag{27}$$

This formula is consistent with the empirical data of Table 4–2 in [34], in the sense that the increasing flame length accelerates the fire spreading.

Inserting (27) into (23) one gets the formula for the ROS in terms of the flame length:

$$V_{\rm ROS} = \frac{\rho c_{\rm p} T_{\rm a}}{\varphi R} \sqrt{2g}\, L_{\rm f_0}^{3/2} (1 + \phi_{\rm wind} + \phi_{\rm slope}) = \frac{\rho c_{\rm p} T_{\rm a}}{\varphi R} \sqrt{2g}\, L_{\rm f}^{3/2} \,. \tag{28}$$

A number of experimental analysis displays a power-law formula of the form

$$L_{\rm f} = \beta_0\, I_{\rm f}^{\beta_1} \,, \tag{29}$$

where β_0 and β_1 are two positive parameters. Formulation (29) is widely used by many authors [3], and the values of β_0 and β_1 emerge to be very scattered, hence formula (24) here derived can be used to provide a theoretical insight for helping in reducing and clarifying such variability.

In particular, by setting

$$\beta_0 = \left[\frac{1}{2g(\rho c_{\rm p} T_{\rm a})^2} \right]^{1/3} \,, \tag{30}$$

from comparison of (24) and (29) follows $\beta_1 = 2/3$, that is dimensionally correct and in agreement with previous empirical and theoretical results [1, 3, 39].

The proposed full derivation of the flame height h – namely (12) plus (21) – and the related flame length L_{f}, can be refined by including the proper environmental characterisation (ρ, c_{p}, T_{a}, $|w|$ and U) through formulae (12) and (21, 17), the vegetation properties (φ, R and V_{ROS_0}) through formula (26), and the configuration parameters (T_{g}, T_{t}, ϕ_{wind} and ϕ_{slope}) through (21) and (24). Moreover, with reference to the phenomenological formula (29), the $\beta_1 = 2/3$ power-law dependence on the fire intensity, as it is observed in the majority of experimental data and expected from dimensional reasons [1], is here obtained for the flame length (11) (or the flame height in the case of steady fire). This suggests that factor β_0 is independent of the fire intensity, according to (30).

In the following section, we discuss how the flame length can be included via the firebrand landing distribution into the fire-spotting model adopted in References [12, 33]. Further, the role of the flame geometry is shown by some numerical tests.

3 Application to the Firebrand Landing Distribution

In view of including the flame length into the model described in References [12, 33], we consider the downwind firebrand landing distance ℓ to be distributed according to a lognormal distribution $q(\ell)$ [16, 17]

$$q(\ell) = \frac{1}{\sqrt{2\pi}\sigma\ell} \exp\left\{-\frac{1}{2}\left[\frac{\ln(\ell/\mu)}{\sigma}\right]^2\right\}, \tag{31}$$

with median μ and mode $\mu e^{-\sigma^2}$, and it holds [16]

$$\mu = H_{\mathrm{max}} \left[\frac{3}{2}\frac{\rho}{\rho_{\mathrm{f}}} C_{\mathrm{d}}\right]^{1/2}, \tag{32}$$

where H_{max} is the maximum loftable height, ρ_{f} is the fuel density and C_{d} is the drag coefficient. In fact, the maximum loftable height H_{max} depends on the fire intensity and atmospheric stability [16, 28]. The detailed study of this parameter and the impact of the atmospheric stability conditions are provided in Reference [12].

In order to include into the fire-spotting model, the flame length through the phenomenological formula (29), the maximum travel distance for a spherical firebrand is written in the following form [37]:

$$\ell_{\mathrm{max}} = H_{\mathrm{max}} \left\{\beta_2 \tan\theta + \left[\frac{3}{2}\frac{\rho}{\rho_{\mathrm{f}}} C_{\mathrm{d}}\mathrm{Fr}\right]^{1/2}\right\}, \tag{33}$$

where $\beta_2 = 0.7$ is a correction factor and $\mathrm{Fr} = U^2/(rg)$ is the Froude number. Note that the angle of the flame can be estimated by the empirical correlation $\tan\theta = 1.35\, U\, (gL_\mathrm{f})^{-1/2}$ [37], that, when applied into the present formulation, gives:

$$\cos\theta = \frac{T_\mathrm{a}(T_\mathrm{g} - T_\mathrm{t})}{(T_\mathrm{t} - T_\mathrm{a})(T_\mathrm{g} - T_\mathrm{a})}\kappa^{1/3} = \left[\frac{gL_\mathrm{f}}{gL_\mathrm{f} + (1.35\, U)^2}\right]^{1/2} . \tag{34}$$

The maximum landing distance can be represented by a certain pth percentile of the lognormal distribution [17], such that

$$\ell_{\max} = \mu \exp(z_\mathrm{p}\sigma) . \tag{35}$$

Thus, from (33) and (35) it holds

$$\sigma = \frac{1}{z_\mathrm{p}} \ln\left\{\mathrm{Fr}^{1/2} + \beta_3\left[\frac{2}{3}\frac{\rho_\mathrm{f}}{\rho}\frac{U^2}{C_\mathrm{d} g L_\mathrm{f}}\right]^{1/2}\right\} , \tag{36}$$

where $\beta_3 = 1.35 \cdot \beta_2 = 0.945$ is a correcting factor [37], and the flame length L_f is defined by formula (29).

Hence from formula (36), the phenomenology reproduced by the present parametrisation is that for increasing flame length L_f the parameter σ decreases and the mode of the lognormal moves towards larger value of the landing distance of the firebrand with the effect of increasing the probability for generating independent separate fires far from the main fire. This effect is studied in the following Section through some test cases performed with the wildfire propagation model described and used for simulations in References [12, 33], but now parameter σ is implemented according to formula (36).

4 Results and Discussions

Here, we study the effects of the flame geometry on fire-spotting. For this purpose, on the basis of some experimental measurements of the flame length, we simulate simple test cases with flat terrain and constant wind. In particular, the impact of the flame geometry on fire-spotting is investigated through the modelling approach described in Reference [12], which means that the flame length is included into the parametrisation of the lognormal distribution of the firebrand landing distance (31) by using formula (36) for the parameter σ.

Briefly, the code used to simulate the fire-front motion is based on the level-set method (LSM) [27], and, at the post-processing stage, the fire-front is then distributed accordingly to the PDF corresponding to the sum of the random fluctuations due to the turbulent heat transport, namely a bi-variate Gaussian density with diffusion coefficient D, and due to the fire-spotting, namely the lognormal (31)

Table 1 Empirical parameters of the flame length-fire intensity relation (29): $L_f = \beta_0\, I_f^{\beta_1}$

References	β_0	β_1
Byram, 1959 [7]	0.0775	0.46
Clark (head fire), 1983 [11]	0.000722	0.99
van Wilgen, 1986 [35]	0.0075	0.46
Fons, 1963 [15]	0.127	2/3
Anderson et al. (Douglas-fir slash), 1966 [4]	0.0447	2/3
Wang, 2011 [37]	0.026445	2/3
Butler, 2004 [6]	0.0175	2/3

with the parametrisation proposed above. Thus, by varying the values of the parameters, different values of σ are obtained.

In particular, with reference to phenomenological formula (29), we study the role of the flame length on fire-spotting by using some experimental estimations of parameters β_0 and β_1, see Table 1. The relationship between flame length and fire intensity may vary because of a different vegetation, this explains the variety of parameters β_0 and β_1 in (29) from the empirical data. In this sense, formula (36) allows also to adjust the fire-spotting model to different vegetation and environmental conditions.

A study on the effect of varying σ in the considered simulation framework has been already discussed in References [12, 17]. The simulations show that σ is an influential factor in the mechanism of generating secondary fires and in their merging with the primary fire, resulting in a complex front-shape with timescales involving σ together with other characteristic features of the process. Concerning the relation between the flame length and the fire intensity, the behaviour of parameter σ with respect to fire intensity in the case of different values of parameters is reported in Fig. 1. This plot shows that for a high enough fire intensity, the standard deviation of the firebrand landing distribution σ is approaching a constant value discovering very slight dependence on fire intensity. From the other side, different power-law formulae of the flame length, which take into account also environmental factors, lead to a quite wide range of possible values of σ.

From our theoretical analysis, the value $\beta_1 = 2/3$ emerges, hence, in order to study the role of the flame length in generating secondary fires, we consider from Table 1 such cases only. The other parameters are set as follows: wind speed $U = 4.47\,\mathrm{ms}^{-1}$, fire intensity $I_f = 20\,\mathrm{MW}^{-1}$ and diffusion coefficient $D = 0.4238\,\mathrm{m}^2\mathrm{s}^{-1}$. Since the flame length does not affect parameter μ, the value $\mu = 8.419$ obtained by the chosen set-up and this value is fixed over all the simulations. The simulated burning areas, at $t = 119$ min, are reported in Figs. 2, 3, 4, and 5 and different fire behaviours are observed.

Actually, it emerges that with larger flame length L_f, that is, small σ, the distribution of the landing distance of the firebrands (31) displays a larger mode that generates long-distance spotting. Hence, the primary fire generates far-away secondary fires that, in turn, rapidly generate further spotting such that the merging results to be slower than the ignition by fire spotting. The final pattern results in

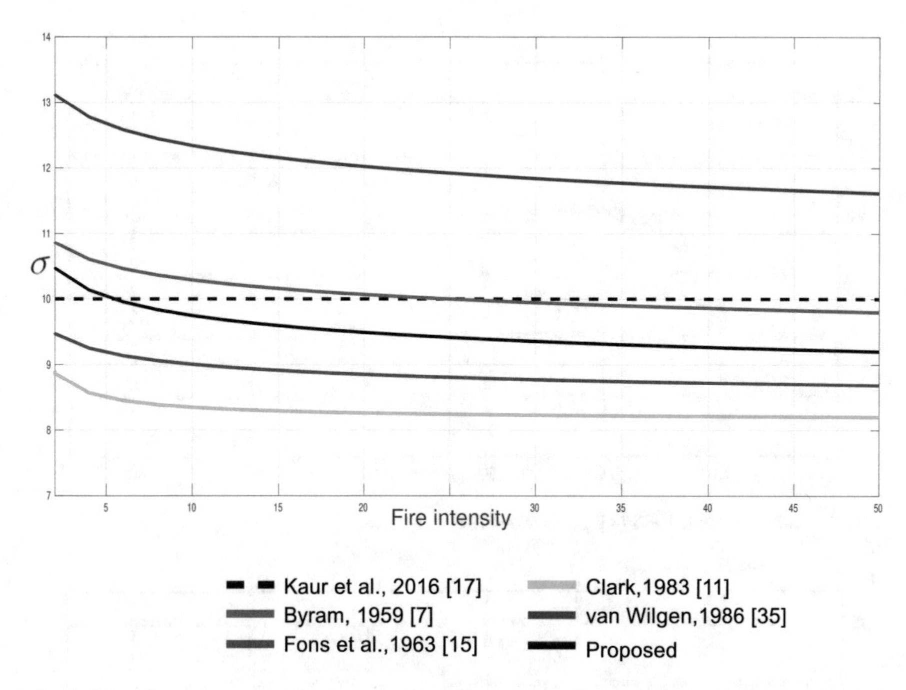

Fig. 1 Plot of parameter σ *vs* fire intensity for different power-law empirical relations, see Table 1 for details

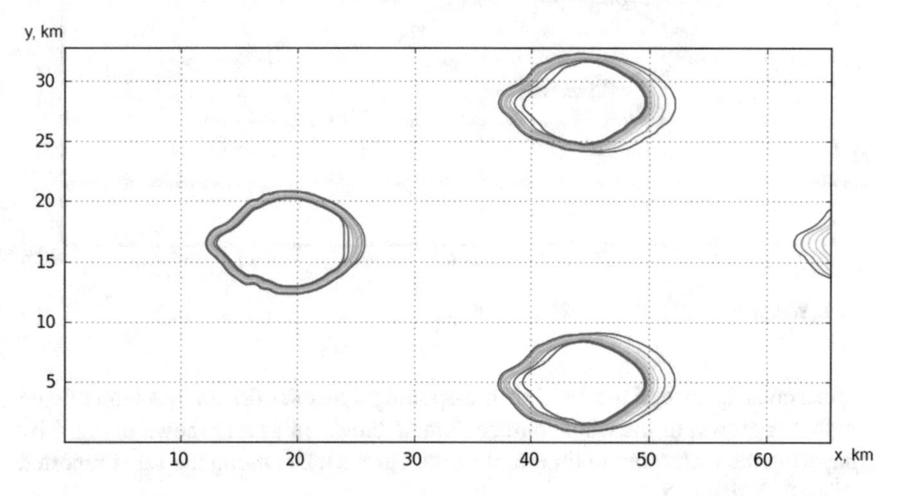

Fig. 2 Fons, 1963 [15]: $\beta_0 = 0.1270$, $\sigma = 5.846$

many independent fires, see Fig. 2. On the contrary, when short-distance spotting takes place, the primary fire rapidly merges with the secondary fires and a unique cumulative burning zone is observed, see Figs. 3, 4, and 5. This is consistent with real fires with many types of fuel and with any fire intensity [31].

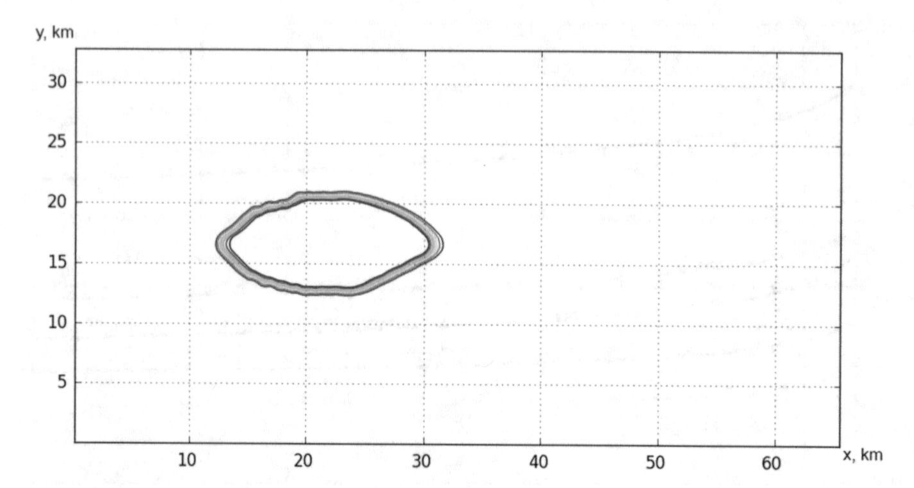

Fig. 3 Anderson et al. 1963 [4]: $\beta_0 = 0.0447$, $\sigma = 6.191$

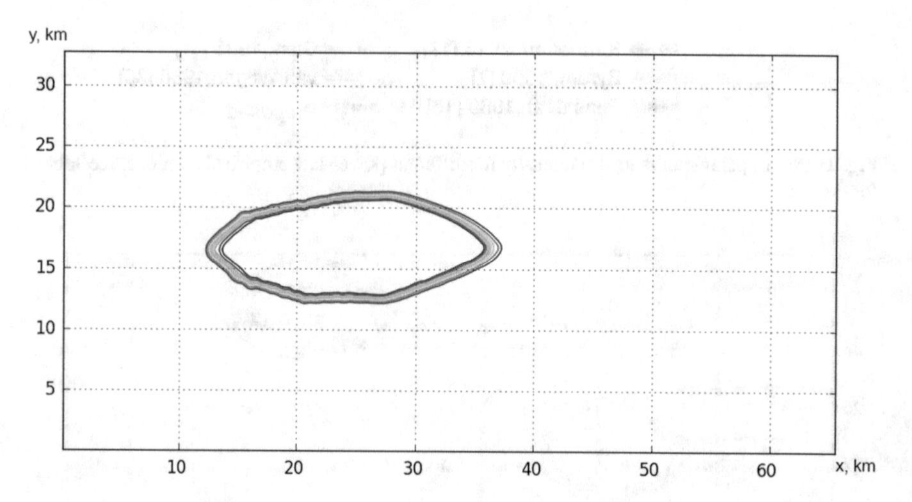

Fig. 4 Wang, 2011 [37]: $\beta_0 = 0.0264$, $\sigma = 6.415$

As a consequence of the multiple fire-spotting dynamics driven by a larger flame length, the growth of the whole burned domain blows off as it is shown in Fig. 6 by comparing the increasing in time of the total burned area among the cases reported in Figs. 2, 3, 4, and 5.

These numerical examples show that the flame geometry, in particular the flame length, is an important factor for the firebrand landing distribution in the sense that fluctuations on this parameter may significantly change the behaviour of the fire-spotting and consequently of the front propagation. In order to validate or reject through the present model these empirical formulae of the flame length-fire intensity interdependences, the role of some environmental and system characteristics

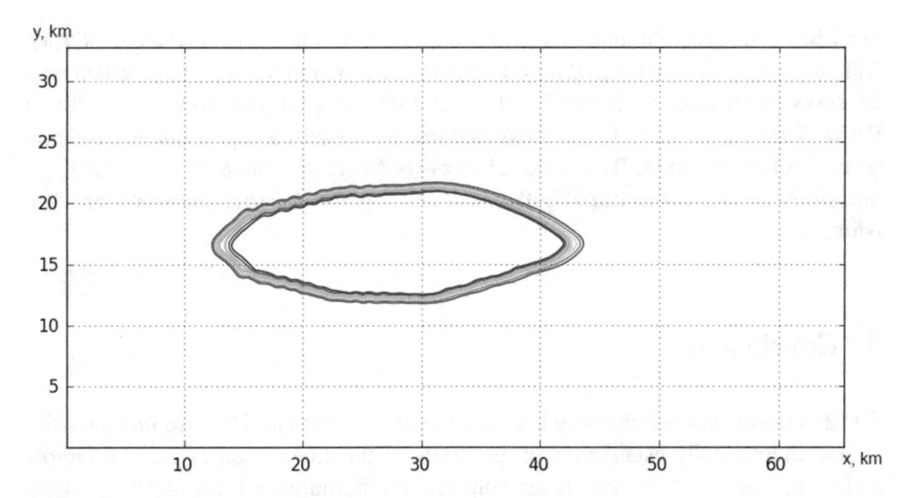

Fig. 5 Butler, 2004 [6]: $\beta_0 = 0.0175$, $\sigma = 6.615$

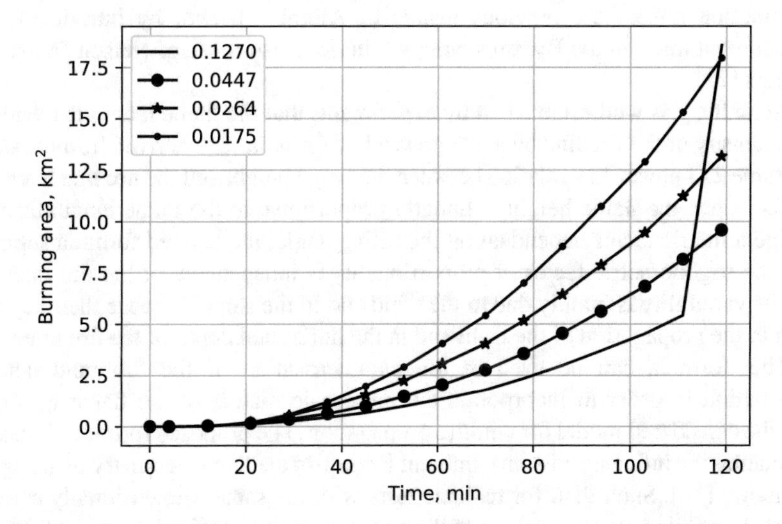

Fig. 6 Increasing in time of the burning area for fixed $\beta_1 = 2/3$ and some empirical values of β_0: $\beta_0 = 0.1270$ [15]; $\beta_0 = 0.0447$ [4]; $\beta_0 = 0.0264$ [37]; $\beta_0 = 0.0175$ [6]. The extreme increasing of burning area in the case with $\beta_0 = 0.1270$ is caused by the multiple fire-spotting dynamics that leads to many independent fires (see Fig. 2), while for the rest of the cases the growth of the burning area is controlled by a rapid merging of secondary fires that leads to a unique cumulative fire (see Figs. 3, 4, and 5).

provided by β_0 according to (30) of by the angle of flame surface by (34) must be also included. In other words, parameters of the flame should be coherent with the system configuration in order to represent correctly the fire behaviour.

The simulations were performed with the code LSFire+, written in C and Fortran, where the model here proposed acts as a post-processing routine at each time step in a LSM code [10]. Simulations have been run on the cluster HYPATIA at BCAM, Bilbao (Basque Country-Spain), by using OpenMP shared memory parallelism over 24 cores of an Intel(R) Xeon(R) CPU E5-2680 v3 2.50 GHz node with 128 GB RAM. Running the code for 45 simulated minutes required approximately 140 min of computational time. The code LSFire+ is freely available at the official git repository of BCAM at https://gitlab.bcamath.org/atrucchia/randomfront-wrfsfire-lsfire.

5 Conclusions

The $2/3$ power-law relationship between the flame length and the fire line intensity is here theoretically established on the basis of the energy conservation principle and of the energy flow rate in the convection column above the fire line, taking into account the presence of the wind and the terrain slope if needed. This new formulation refines the previous results of Albini [1], and, by introducing the entrainment through the Byram's energy criterion, also those of Nelson Jr. and co-authors [21].

Actually, it is well established by experiments that the flame length is related to the intensity of the fire line by a $2/3$ power-law formula. The derived formula states the same $2/3$ power-law relation between the flame height and the fire line intensity. Hence, since the flame height is linearly proportional to the flame length through a trigonometric factor dependent on the tilting angle, the derived formula suggests that the trigonometric factor of proportionality is independent of the fire intensity and its variability is mainly due to the wind and to the slope, because their key role both in the propagation of the front and in the horizontal depth of the fire line.

This formula can be used in the parametrisation of the firebrand landing distribution in order to incorporate the mesoscale factors of the flame geometry into the considered model for wildfire propagation. The proposed formula allows for estimating the influence of some ambient factors on the flame geometry by using the definition (30). Such that, for realistic tests with the same (dimensionally correct) power-law factor, smaller values of the proportionality coefficient represent higher ambient temperature or vegetation with higher specific heat. Hence, for different types of fuel the flame length relates to the fire intensity through different coefficient of proportionality. Thus, the proposed formula allows to specify the fire-spotting model in accordance with the type of fuel in each particular case.

It is found that the flame geometry is significant for the fire-spotting, thus, further study of this phenomenon is required. It is well known that the flame length is dependent also on the slope of the terrain, thus the topography of the surround can be introduced into the front propagation model. Such fire-spotting models may significantly improve the prediction capabilities of the existing wildfire propagation models.

Acknowledgments This research is supported by the Basque Government through the BERC 2014-2017 and the BERC 2018-2021 programs and by Spanish Ministry of Economy and Competitiveness MINECO through BCAM Severo Ochoa excellence accreditations SEV-2013-0323 and SEV-2017-0718 and through project MTM2016-76016-R 'MIP' and by the PhD grant 'La Caixa 2014'.

References

1. Albini, F.A.: A model for the wind-blown flame from a line fire. Combust. Flame **43**, 155–174 (1981)
2. Alexander, M.E.: Calculating and interpreting forest fire intensities. Can. J. Bot. **60**, 349–357 (1982)
3. Alexander, M.E., Cruz, M.G.: Interdependencies between flame length and firefire intensity in predicting crown fire initiation and crown scorch height. Int. J. Wildland Fire **21**, 95–113 (2012)
4. Anderson, H.E., Brackebusch, A.P., Mutch, R.W., Rothermel, R.C.: Mechanisms of fire spread research, Progress Report No. 2. Research Paper INT-RP-28, USDA Forest Service, Intermountain Forest and Range Experiment Station, Ogden (1966)
5. Andrews, P.L.: The Rothermel surface fire spread model and associated developments: a comprehensive explanation. General Technical Reports RMRS-GTR-371. U.S. Department of Agriculture, Forest Service, Rocky Mountain Research Station, Fort Collins (2018)
6. Butler, B.W., Finney, M.A., Andrews, P.L., Albini, F.A.: A radiation-driven model of crown fire spread. Can. J. For. Res. **34**, 1588–1599 (2004)
7. Byram, G.M.: Combustion of forest fuels. In: Davis, K.P. (ed.) Forest Fire: Control and Use, pp. 61–89. McGraw Hill, New York (1959)
8. Byram, G.M.: Forest fire behavior. In: Davis, K.P. (ed.) Forest Fire: Control and Use, pp. 90–123. McGraw Hill, New York (1959)
9. Campbell, M.J., Dennison, P.E., Butler, B.W.: Safe separation distance score: a new metric for evaluating wildland firefighter safety zones using lidar. Int. J. Geogr. Inf. Sci. **31**(7), 1448–1466 (2016)
10. Chu, K.T., Prodanović, M.: Level set method library (LSMLIB). http://ktchu.serendipityresearch.org/software/lsmlib/ (2009)
11. Clark, R.G.: Threshold requirements for fire spread in grassland fuels. Ph.D. thesis, Texas Tech University, Lubbock (1983)
12. Egorova, V.N., Trucchia, A., Pagnini, G.: Fire-spotting generated fires. Part I: the role of atmospheric stability. Appl. Math. Model. **84**, 590–609 (2020). https://doi.org/10.1016/j.apm.2019.02.010
13. Fernandez-Pello, A.C.: Wildland fire spot ignition by sparks and firebrands. Fire Safety J. **91**, 2–10 (2017)
14. Ferragut, L., Asensio, M.I., Cascón, J.M., Prieto, D.: A wildland fire physical model well suited to data assimilation. Pure Appl. Geophys. **172**, 121–139 (2015)
15. Fons, W.L., Clements, H.B., George, P.M.: Scale effects on propagation rate of laboratory crib fires. Symp. Int. Combust. Proc. **9**, 860–866 (1963)
16. Kaur, I., Pagnini, G.: Fire-spotting modelling and parametrisation for wild-land fires. In: Sauvage, S., Sánchez-Pérez, J.M., Rizzoli, A.E. (eds.) Proceedings of the 8th International Congress on Environmental Modelling and Software (iEMSs2016); Toulouse, 10–14 July (2016), pp. 384–391 (2016). ISBN: 978-88-9035-745-9
17. Kaur, I., Mentrelli, A., Bosseur, F., Filippi, J.-B., Pagnini, G. Turbulence and fire-spotting effects into wild-land fire simulators. Commun. Nonlinear Sci. Numer. Simul. **39**, 300–320 (2016)

18. Marcelli, T., Balbi, J.H., Moretti, B., Rossi, J.L., Chatelon, F.J.: Flame height model of a spreading surface fire. In: Proceedings of the 7th Mediterranean Combustion Symposium MCS7, Cagliari, 11–15 Sept (2011). ISBN: 978-88-88104-12-6
19. Martin, J., Hillen, T.: The spotting distribution of wildfires. Appl. Sci. **6**(6), 177–210 (2016)
20. Nelson, R.M. Jr.: Byram's energy criterion for wildland fires: units and equations. Research Note INT-415, Intermountain Research Station, Forest Service (1993)
21. Nelson, R.M. Jr., Butler, B.W., Weise, D.R.: Entrainment regimes and flame characteristics of wildland fires. Int. J. Wildland Fire **21**, 127–140 (2012)
22. Nmira, F., Consalvi, J.L., Boulet, P., Porterie, B.: Numerical study of wind effects on the characteristics of flames from non-propagating vegetation fires. Fire Safety J. **45**(2), 129–141 (2010)
23. NWCG: S-290 intermediate wildland fire behavior course. Unit 12: gauging fire behavior and guiding fireline decisions
24. Pagnini, G., Mentrelli, A.: Modelling wildland fire propagation by tracking random fronts. Nat. Hazards Earth Syst. Sci. **14**, 2249–2263 (2014)
25. Rossi, J.L., Chetehouna, K., Collin, A., Moretti, B., Balbi, J.H.: Simplified flame models and prediction of the thermal radiation emitted by a flame front in an outdoor fire. Combust. Sci. Technol. **182**, 1457–1477 (2010)
26. Rothermel, R.C.: A mathematical model for predicting fire spread in wildland fires. Research Paper INT-115, USDA Forest Service, Intermountain Forest and Range Experiment Station, Ogden (1972). Available at: http://www.treesearch.fs.fed.us/pubs/32533
27. Sethian, J.A., Smereka, P.: Level set methods for fluid interfaces. Ann. Rev. Fluid Mech. **35**, 341–372 (2003)
28. Sofiev, M., Ermakova, T., Vankevich, R.: Evaluation of the smoke-injection height from wildland fires using remote-sensing data. Atmos. Chem. Phys. **12**(4), 1995–2006 (2012)
29. Sullivan, A.L.: Convective Froude number and Byram's energy criterion of Australian experimental grassland fires. Proc. Combust. Inst. **31**, 2557–2564 (2007)
30. Sullivan, A.L.: Inside the inferno: fundamental processes of wildland fire behaviour. Part 1: combustion chemistry and heat release. Curr. Forest. Rep. **3**, 132–149 (2017)
31. Sullivan, A.L.: Inside the inferno: fundamental processes of wildland fire behaviour. Part 2: heat transfer and interaction. Curr. Forest. Rep. **3**, 150–171 (2017)
32. Thomas, P.: The size of flames from natural fires. Symp. Int. Combust. Proc. **9**, 844–859 (1963)
33. Trucchia, A., Egorova, V., Butenko, A., Kaur, I., Pagnini, G.: RandomFront 2.3: a physical parametrisation of fire-spotting for operational fire spread models – implementation in WRF-SFIRE and response analysis with LSFire+. Geosci. Model Dev. **12**, 69–87 (2019)
34. Vaillant, N.M., Ager, A.A., Anderson, J., Miller, L.: ArcFuels user guide and tutorial: for use with ArcGIS 9. General Technical Report PNW-GTR-877, U.S. Department of Agriculture, Forest Service, Pacific Northwest Research Station (2013)
35. van Wilgen, B.W.: A simple relationship for estimating the intensity of fires in natural vegetation. S. Afr. J. Bot. **52**, 384–385 (1986)
36. Viegas, D.X. (ed.): Forest Fire Research & Wildland Fire Safety. Millpress, Rotterdam (2002)
37. Wang, H.-H.: Analysis on downwind distribution of firebrands sourced from a wildland fire. Fire Technol. **47**(2), 321–340 (2011)
38. Weise, D.R., Fletcher, T.H., Zhou, S.M.X., Sun, L.: Fire spread in chaparral: comparison of data with flame-mass loss relationships. In: Proceedings of the 8th International Symposium on Scale Modeling (ISSM-8), 12–14 Sept 2017, Portland (2017)
39. Weise, D.R., Fletcher, T.H., Cole, W., Mahalingam, S., Zhou, X., Sun, L., Li, J.: Fire behavior in chaparral – evaluating flame models with laboratory data. Combust. Flame **191**, 500–512 (2018)
40. Wilson, A.A.G., Ferguson, I.S.: Predicting the probability of house survival during bushfires. J. Environ. Manage. **23**, 259–270 (1986)

Wind Shear Forecast in GCLP and GCTS Airports

David Suárez-Molina and Juan Carlos Suárez González

Abstract Low-Level Wind Shear (LLWS) is one of the most critical aviation hazards. Detecting it accurately must be the main objective to guarantee flight safety. Terrain-induced wind shear at Tenerife South (GCTS) and Gran Canaria airports (GCLP) could be hazardous to the landing and departing aircraft. This paper shows an experimental wind shear alert system based on two different methodologies. Both of them have been developed from u and v wind components of Harmonie-Arome Model.

1 Introduction

Wind shear is a change in wind speed and/or direction over a short distance. It can occur either horizontally, vertically or both and is most often associated with strong temperature inversions or density gradients. The meteorological phenomena and atmospheric conditions that are hazards with the potential of causing aircraft accidents are well known [6]. Detecting it accurately, as well as being able to alert users, should be the main objective to ensure flight safety [5]. The causes of wind shear are very well known. Convective weather with first gusts, downdrafts, microbursts, and gravity waves are the most significant forms of windshear. Terrain features like mountains, gullies, or other topography cause wind flows to change over short distances. Man-made obstacles, like a large hangar beside the runway, create a changing wind pattern. Fronts and storms can create vertical shearing in the atmosphere close to the ground. Wind shear from each of these causes has made an impact on some airplane in the last few decades [7]. Wind shear can occur at any level of the atmosphere, nevertheless Low-Level Wind Shear (LLWS) can be very dangerous. During the 1974–1999 period, over 650 deaths took place in commercial aviation alone due to wind shears [7].

D. Suárez-Molina (✉) · J. C. S. González
State Meteorological Agency (AEMET), Delegación Territorial de AEMET en Canarias,
Las Palmas de Gran Canaria, Spain
e-mail: dsuarezm@aemet.es

M. I. Asensio et al. (eds.), *Applied Mathematics for Environmental Problems*,
SEMA SIMAI Springer Series 6, https://doi.org/10.1007/978-3-030-61795-0_3

Several authors have analyzed the relationship between aviation and weather under a meteorological perspective. Ágústsson and Ólafsson studied a case of severe turbulence, caused by orography, affecting an aircraft when flying over the southeastern coast of Iceland [1]. They used numerical simulations to describe a downslope windstorm at the ground associated with amplified lee waves and rotor aloft. Strong shear turbulence was simulated at the interface of the lee wave and the rotor, which produced severe turbulence. However, to the authors' knowledge, there are only a few studies carried out in Canary airports.

This study is organized as follows. A brief description of the data and methods are incorporated in Sect. 2. Subsequently, Sect. 3 exposes the main results of this research and the discussion, including an examination of case studies. Finally, the main conclusions of this paper are provided in Sect. 4.

2 Data and Methods

2.1 Study Area

The Canary archipelago (Fig. 1), which is located west of North Africa, consists of seven islands with a total area of about 7200 km^2, 1100 km away from the Spanish mainland. Reaching from 27°37′ to 29°25′N and from 18°10′ to 13°20′W, all islands belong to the subtropical zone. This research is centered in Gran Canaria (code ICAO: GCLP) and Tenerife Sur (code ICAO: GCTS) airports.

GCLP is located in the east of Gran Canaria at 24 m AMSL (above mean sea level). GCTS is located in the south of Tenerife at 64 m AMSL. The runways have oriented to the Trade Wind from the north-east. In 2018, according to data from the official AENA website the airports handled more than 24 million of passengers.

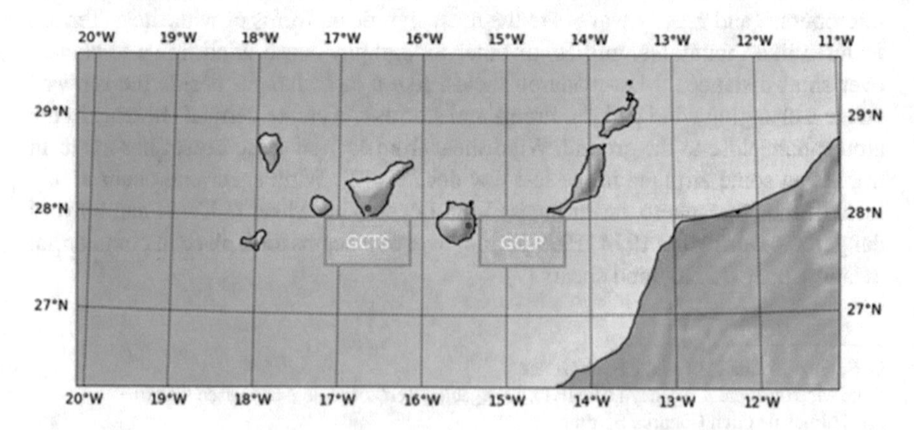

Fig. 1 Study area, The Canary Islands. The red dots highlight GCLP (Gran Canaria airport) and GCTS (Tenerife Sur airport)

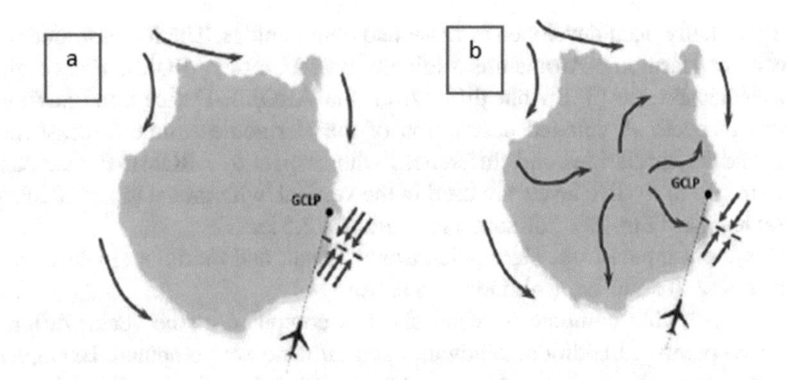

Fig. 2 LLWS conceptual model in the Canary airports. Red line highlights the convergence line. (**a**) Low height temperature inversion. (**b**) High height temperature inversion

During the 2015–2018 period, more than 15.000 LLWS ATIS (Automatic Terminal Information Service) message were analyzed. The statistical analysis shows that the 85% of LLWS cases were terrain induced. On the other hand, only 9% of LLWS cases were related to thunderstorms or fronts.

2.2 *LLWS Conceptual Model*

The majority of significant LLWS events at GCLP and GCTS could be associated with terrain disruption of airflow. Gran Canaria and Tenerife present a complex orography in the center of the islands and deep ravines. Winds blowing across the mountains from the northwest or southwest would become disturbed by terrain and might bring about significant windshear or turbulence downstream. By the other hand, the flow can surround the islands and convergence line can be originated close to airports. The temperature inversion height will determine if the flow goes over the mountains or on the contrary surround the islands (Fig. 2).

The presence of terrain will lead to faster airstreams across the gaps and slower wind speeds directly behind the mountain. An aircraft flying through the region will experience sudden changes in headwind and hence windshear.

2.3 *Data*

Input data are obtained from the predictions of the Numerical Weather Prediction model. In this research, we used Harmonie-Arome. The non-hydrostatic convection-permitting Harmonie-Arome model is developed in a code cooperation with Météo-France and ALADIN, and builds upon model components that have largely initially been developed in these two communities. The forecast model and analysis of Harmonie-Arome are originally based on the AROME-France model from Météo-France [3, 8], but differ from the AROME-France configuration in various respects. A detailed description of the Harmonie-Arome forecast model setup and its similarities and differences with respect to AROME-France can be found in [2]. Sixty-five levels are used in the vertical, with model top at 10 hPa and lowest level at 12 m. The horizontal resolution is 2.5 km.

As system inputs from Harmonie-Arome, zonal, and meridional wind components (u and v) at different altitudes are used.

From u-v wind components wind shear is computed as the vector difference from two points. In addition, headwind, and tailwind are computed. Examples of how wind shear, headwind, and tailwind are computed can be found in Manual on Low-Level Wind Shear published by ICAO [4].

2.4 Experimental Warning Systems

From wind shear, headwind, and tailwind, two experimental warnings systems have been developed. The results are showed through a visualization application a can be seen in Figs. 3 and 4. These systems will be described below.

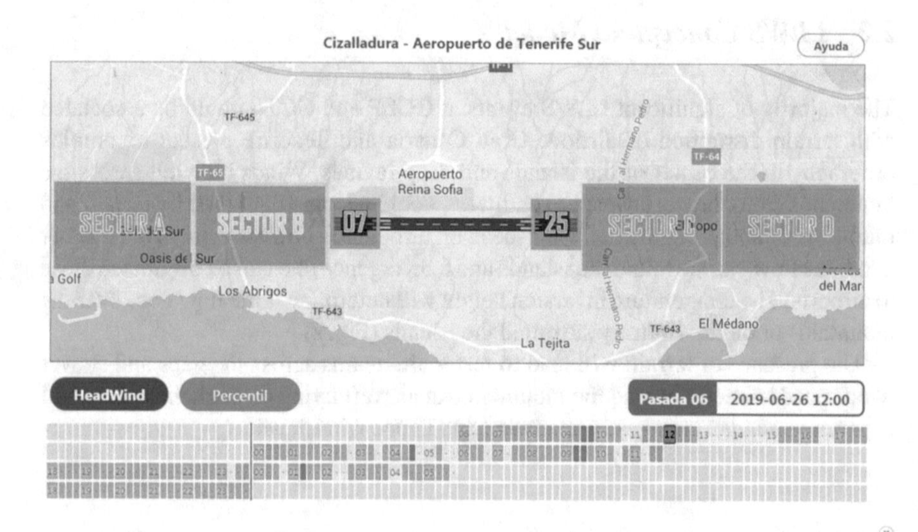

Fig. 3 Headwind experimental warning system

2.4.1 Headwind System

In this system (Fig. 3), headwind is computed for each airport. Then the difference of headwind between two consecutive time steps in different locations is calculated. Time step is 15 min and the locations are showed in Fig. 3. The sector B and sector C are located 1 M from runway at 100 m AMSL. The sector A and sector D are located 2 M from runway at 200 m AMSL. In the horizontal, the nearest model grid point to the different sectors has been selected. In the vertical, we used model outputs in height levels at 100 and 200 m.

Different visual warnings will be issued by the system depending on sustainable change of headwind. So if the differences of headwind between (in a specific sector) two consecutive time steps is between 15 and 20 kn, the systems will give us a yellow warning, an orange warning between 21 and 30 kn, and a red warning above 30 kn.

2.4.2 Percentile System

In this system (Fig. 4), wind shear within each locations is computed as the vector difference. The locations are called APCH XX, APCH YY (where XX and YY are the runway) and RUNWAY. The wind shear computed is compared to established thresholds. The thresholds have been established according to the 90th, 95th and 99th percentiles of a time series previously studied (time series includes from January 1, 2017 to May 31, 2018). When wind shear is between 90th and 95th percentile, yellow visual warning will be issued by the system. While wind shear

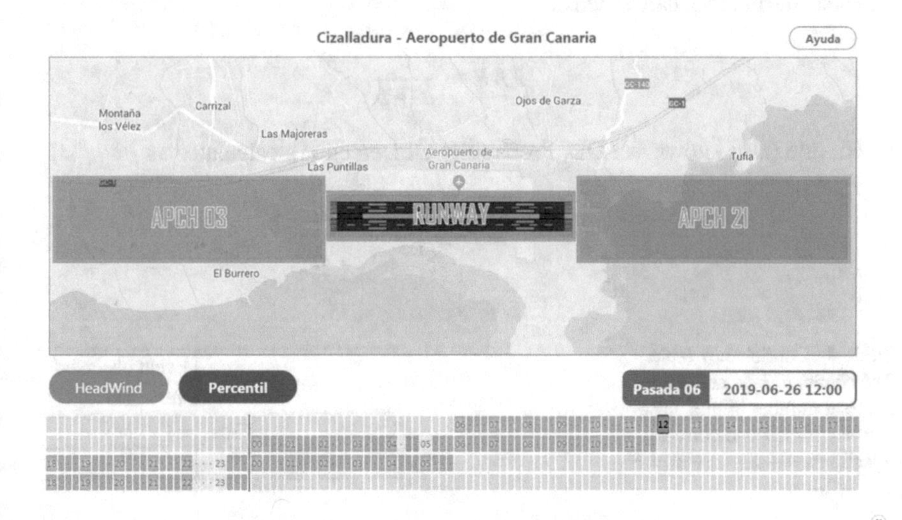

Fig. 4 Percentile experimental warning system

is between 95th and 99th percentile, orange visual warning will be issued by the system. Finally, if 99th percentile is exceeded, red visual warning will be issued by the system.

2.5 Systems Evaluation

The wind shear warnings forecast were compared with the measurements of LLWAS (low level windshear alert system, located in GCTS) and with LLWS ATIS message (in GCLP, because in this airport there is not a LLWAS). Wind shear was regarded as a simple binary event and summarized by 2×2 contingency table (Table 1). The table elements are hits (correct forecast and event), misses (observed but not forecasted event), false alarms (forecast but no observed event), and correct rejections (correct forecast of non-event) [9]. Based on the contingency table different scores were computed:

- Accuracy, calculated as

$$PC = \frac{a+b}{n}$$

- Frequency bias, calculated as

$$FB = \frac{a+b}{a+c}$$

- False alarm ratio, calculated as

$$FAR = \frac{b}{a+b}$$

- Hit rate (also known as POD, Probability Of Detection), calculated as

$$H = \frac{a}{a+c}$$

Table 1 Contingency table. The counts a, b, c, and d are the total number of hits, false alarms, misses, and correct rejections. $a + b + c + d = n$

		Event observed	
		Yes	No
Event forecast	Yes	a	b
	No	c	d

- False alarm rate, calculated as

$$F = \frac{b}{b+d}$$

- True statistic skill, calculated as

$$TSS = ad - \frac{bc}{(a+c)(b+d)}$$

Relative operating characteristic (ROC) plot hit rate (POD) versus false alarm rate (POFD), using a set of increasing probability thresholds (e.g., 0.05, 0.15, 0.25) to make the yes/no decision. The area under the ROC curve is frequently used as a score. ROC shows a perfect score (area = 1) when curve travels from bottom left to top left of diagram, then across to top right of diagram. Diagonal line indicates no skill (area = 0.5).

ROC measures the ability of the forecast to discriminate between two alternative outcomes, thus measuring resolution. It is not sensitive to bias in the forecast, so says nothing about reliability. A biased forecast may still have good resolution and produce a good ROC curve, which means that it may be possible to improve the forecast through calibration. The ROC can thus be considered as a measure of potential usefulness.

The ROC is conditioned on the observations (i.e., given that an event occurred, what was the corresponding forecast?) It is therefore a good companion to the reliability diagram, which is conditioned on the forecasts.

In addition, runway wind shear was compared with the measurements gathered at the airports' automatic meteorological stations. In this case, forecast verification for continuous predictands was used.

Three different error measurements were used: BIAS (the correspondence between the mean forecast and mean observation), mean absolute error (MAE), and root mean square error (RMSE). The mathematical formulas are the following:

$$\text{BIAS} = \frac{1}{N}\sum_{i=1}^{N}(F_i - O_i)$$

$$\text{MAE} = \frac{1}{N}\sum_{i=1}^{N}|F_i - O_i|$$

$$\text{RMSE} = \sqrt{\frac{1}{N}\sum_{i=1}^{N}(F_i - O_i)^2},$$

where F_i is the forecasted wind at location i, O_i is the observed wind at location i, and N is the total number of locations.

3 Results

The main verification results will be showed in this chapter. In order to support the results, one study case has been included as part of the chapter.

3.1 Verification

As mentioned in previously chapter, in order to evaluate the experimental warnings systems, verification was performed using the results over a 3 month period (September 2018 to November 2018).

3.1.1 Discrete Verification

Methods for dichotomous forecast and observed events were applied in order to compare LLWS forecast. Table 3 shows the scores for GCLP and GCTS by airports computed from the contingency tables (Table 2).

PC score indicates what fraction of the forecasts were correct. Perfect score is 1. According to the results in Table 3, PC is close to or above 0.8 (except GCLP headwind system) indicating that around 80% of all forecasts were correct. It can be misleading since it is heavily influenced by the most common category.

FB score measures the ratio of the frequency of forecast events to the frequency of observed events. It indicates whether the forecast system has a tendency to underforecast (FB < 1) or overforecast (FB > 1) events. It does not measure how well the forecast corresponds to the observations, only measures relative frequencies. The results indicate that the systems have a tendency to overforecast.

Table 2 Contingency table

GCLP

Headwind system				Percentile system			
		Event observed				Event observed	
		Yes	No			Yes	No
Event forecast	Yes	193	4050	Event forecast	Yes	98	929
	No	140	3490		No	235	6460

GCTS

Headwind system				Percentile system			
		Event observed				Event observed	
		Yes	No			Yes	No
Event forecast	Yes	120	1327	Event forecast	Yes	223	2474
	No	432	12,003		No	329	10,858

Table 3 Scores computed for GCLP and GCTS by system

	GCLP		GCTS	
	Headwind system	Percentile system	Headwind system	Percentile system
PC	0.47	0.85	0.87	0.80
FB	12.74	3.08	2.62	4.89
FAR	0.95	0.90	0.92	0.92
H	0.58	0.29	0.22	0.40
F	0.54	0.12	0.10	0.19
TSS	0.04	0.17	0.12	0.22

FAR score is very high for both systems (perfect score is 0), this score indicates that above 90% of the forecast windshear events, windshear was not observed.

H indicates what fraction of the observed 'yes' events were correctly forecast. GCLP headwind system shows the best H score followed by GCTS percentile system.

F is sensitive to false alarms, but ignores misses. Can be artificially improved by issuing fewer 'yes' forecasts to reduce the number of false alarms. Perfect score is 0 and in general the systems shows good results (except for GCLP headwind system).

TSS shows how well did the forecast separate the 'yes' events from the 'no' events. Perfect score is 1. Table 3 indicates that the systems are slightly better than random forecasts.

3.1.2 Continuous Verification

Verification of continuous variables is included in this section. It measures how the values of the forecasts differ from the values of the observations. Runway wind shear was compared with the measurements gathered at the airports' automatic meteorological stations. Table 4 shows BIAS, MAE, and RMSE for GCLP and GCTS runways. BIAS values are negative but close to 0 (perfect score is 0). That is, mean system has a tendency to under forecast. MAE values indicate that the average magnitude of the forecast errors is around 1.5 and RMSE values show that the average magnitude of the forecast errors are above 2.3 kn.

Figure 5 shows runway wind shear for GCTS during the September-November 2018 period. Red line corresponds to forecast and gray line is observed data. In spite of negative bias, the forecasting data are quite similar to observed data. Negative bias is clearly shown in scatter plots (Fig. 6).

Table 4 Scores computed for runway windshear in GCLP and GCTS

	BIAS	MAE	RMSE
GCLP	−0.676	1.521	2.314
GCTS	−0.382	1.587	2.449

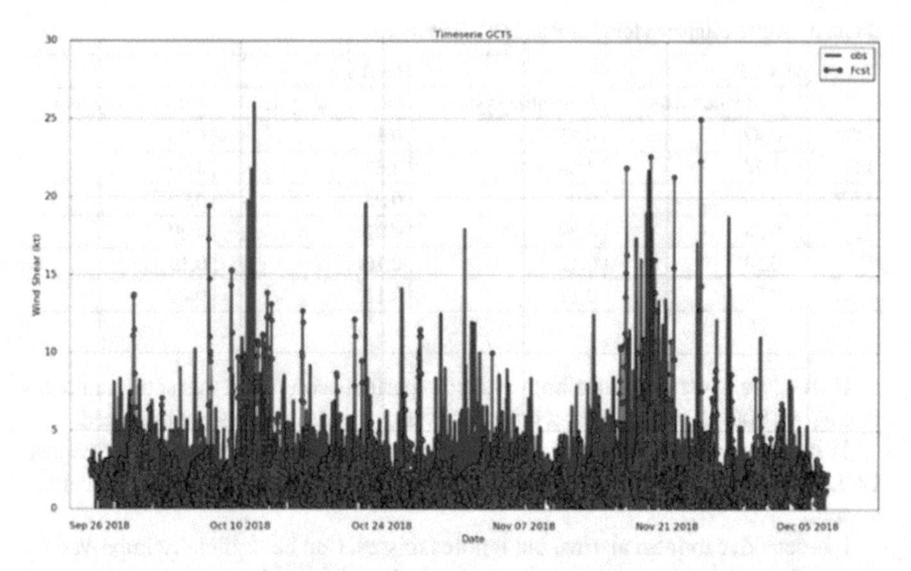

Fig. 5 Runway windshear time series (the graphs correspond to GCTS), of 15 min data. Red line corresponds to forecast and gray line is observed data

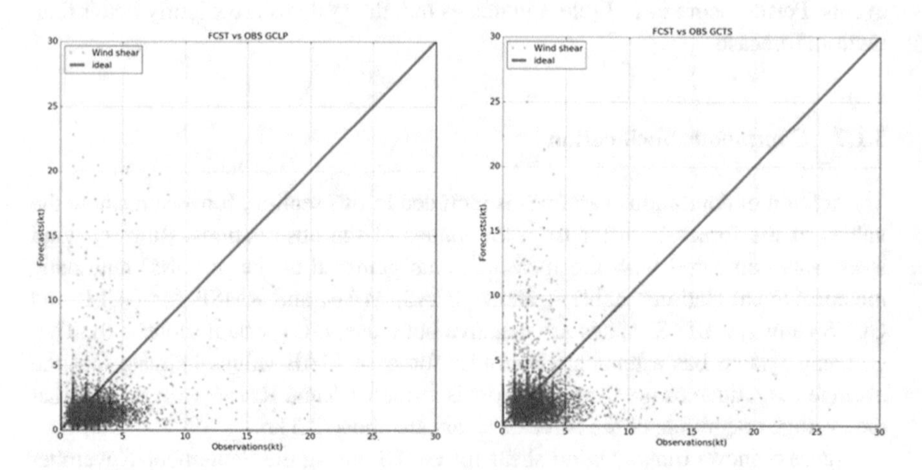

Fig. 6 Scatter plots of runway windshear. Left GCLP airport. Right GCTS airport

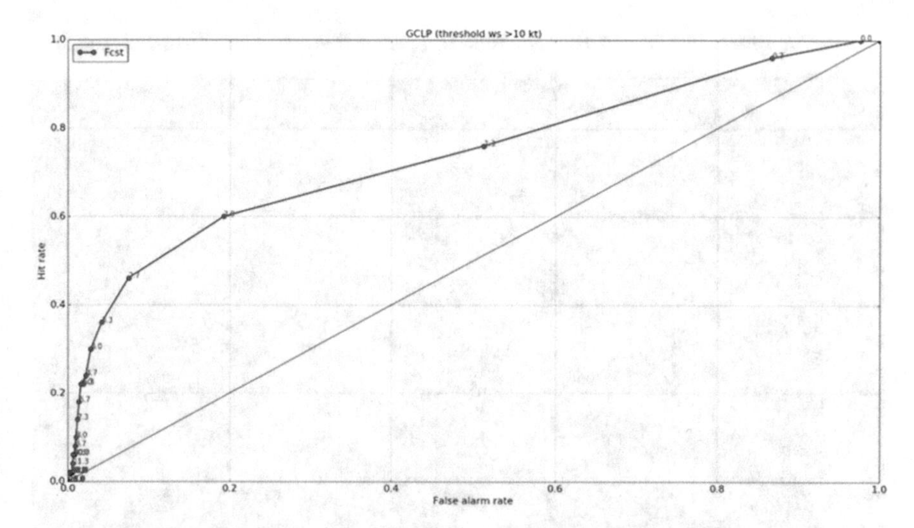

Fig. 7 ROC plot of runway windshear for GCLP. Threshold established was 10 kn

ROC (Fig. 7) measures the ability of the forecast to discriminate between two alternative outcomes, thus measuring resolution. In spite of the biased forecast, it has good resolution and produces a good ROC curve, which means that it may be possible to improve the forecast through calibration.

3.2 Study Case

In this section, a study case will be analyzed. On February 28, 2018, 21 go-arounds took place at GCLP.

The Canary Islands were affected by extra-tropical low pressure system and western winds blew over the islands. Close to the airport GCLP mean wind was around 40 kn (74 km h^{-1}) and the wind gust reached 60 kn (111 km h^{-1}). This situation can be seen in Fig. 8. This figure shows mean wind and wind gust from Harmonie-Arome model over Gran Canaria. GCLP is highlighted in yellow.

In addition, visible image from satellite Meteosat is shown (Fig. 9). In the satellite image it can be appreciated mesoscale convergence line located southwest of Gran Canaria. This line was moving to the northeast affecting to GCLP.

The experimental products showed (Fig. 10) a high probability of tailwind in approach and runway, coinciding with a high number of go around operations.

Fig. 8 Mean wind and wind gust from Harmonie-Arome (Gran Canaria). Arrows are mean wind with colors indicating speed and wind gust are contouring

4 Conclusions

The general conclusions that can be drawn from this study are:

- Statistical analysis has shown that LLWS in the Canary airports is mainly terrain induced.
- Two experimental wind shear alert system has been tested. These systems are based on the forecast of an NWP model.
- In spite of turbulence and therefore wind shear is a microscale phenomenon, mesoscale models, like as Harmonie-Arome, can help to forecast wind shear events.
- Verification methods for dichotomous forecast showed that for GCLP percentile system got better results and for GCTS headwind system was better than percentile system.
- Both systems have a positive frequency bias having a tendency to overforecast.
- On the contrary, continuous verification of runaway wind shear against automatic meteorological station shows a slight tendency to underforecast wind shear. This

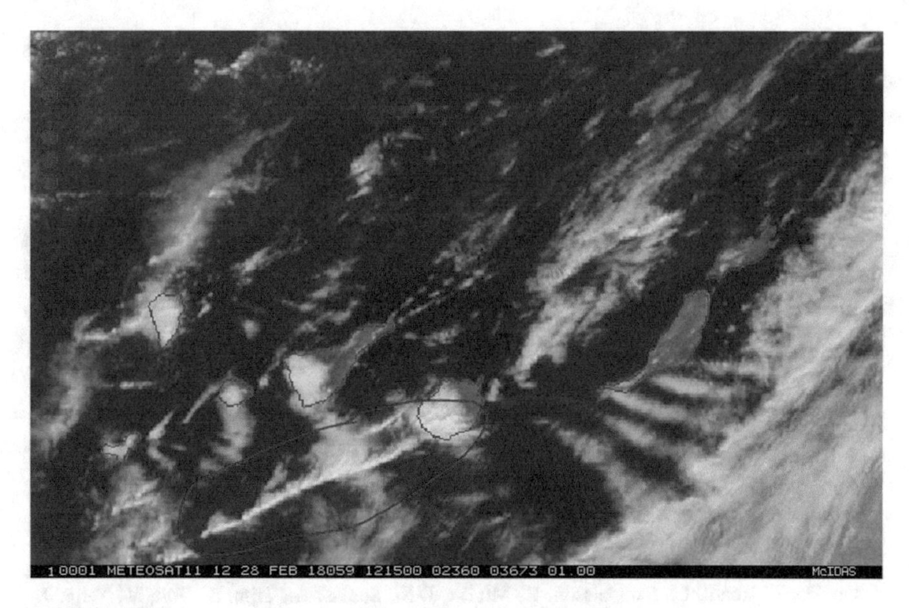

Fig. 9 Visible image from Meteosat satellite (February 28, 2018, Canary Islands)

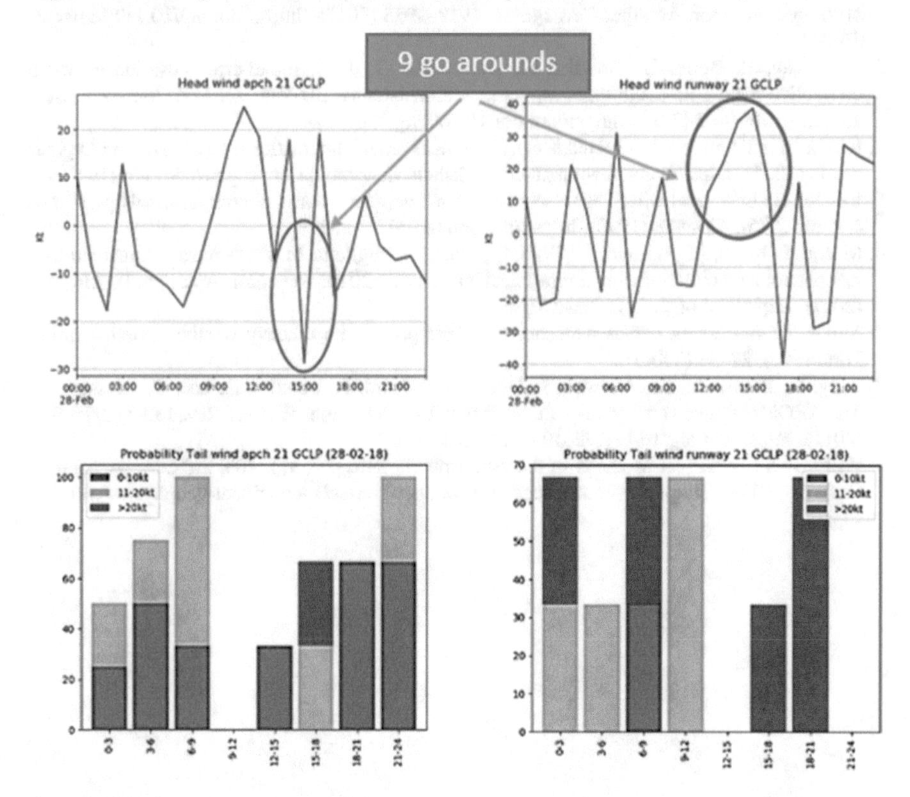

Fig. 10 Above: Headwind time series in the sectors approach 21 and 21 runway of GCLP. Below: Probability tailwind in the sectors approach 21 and 21 runway of GCLP (February 28, 2018)

contradiction between the positive frequency bias from systems and negative bias from runaway wind shear could have several origins:

– *u*/*v* components from NWP model are biased.
– ATIS are subjective messages.
– Thresholds used by systems are low.

- As future work, the authors think that the improvement in the NWP model resolution can help make more accurate forecasts. In addition, ROC curve pointed that it may be possible to improve the forecast through model calibration.

References

1. Ágústsson, H., Ólafsson, H.: Simulations of observed lee waves and rotor turbulence. Mon. Weather Rev. **142**(2), 832–849 (2014). https://doi.org/10.1175/mwr-d-13-00212.1
2. Bengtsson, L., Andrae, U., Aspelien, T., Batrak, Y., Calvo, J., de Rooy, W., Gleeson, E., Hansen-Sass, B., Homleid, M., Hortal, M., Ivarsson, K.I., Lenderink, G., Niemelä, S., Nielsen, K.P., Onvlee, J., Rontu, L., Samuelsson, P., Muñoz, D.S., Subias, A., Tijm, S., Toll, V., Yang, X., Køltzow, M.Ø.: The HARMONIE–AROME model configuration in the ALADIN–HIRLAM NWP system. Mon. Weather Rev. **145**(5), 1919–1935 (2017). https://doi.org/10.1175/mwr-d-16-0417.1
3. Brousseau, P., Berre, L., Bouttier, F., Desroziers, G.: Background-error covariances for a convective-scale data-assimilation system: AROME-france 3D-Var. Q. J. R. Meteorol. Soc. **137**(655), 409–422 (2011). https://doi.org/10.1002/qj.750
4. ICAO: Manual on low level wind shear. Technical report, International Civil Aviation Organization (2005). https://www.skybrary.aero/bookshelf/views/bookDetails.php?bookId=2194
5. Kessler, E.: Low-level windshear alert systems and doppler radar in aircraft terminal operations. J. Aircr. **27**(5), 423–428 (1990). https://doi.org/10.2514/3.25293
6. Mazon, J., Rojas, J.I., Lozano, M., Pino, D., Prats, X., Miglietta, M.M.: Influence of meteorological phenomena on worldwide aircraft accidents, 1967–2010. Meteorol. Appl. **25**(2), 236–245 (2017). https://doi.org/10.1002/met.1686
7. Minor, T.: Judgement versus windshear: an epic poem with a heavy weather moral. Mobil. Forum **9**(3), 27–33 (2000)
8. Seity, Y., Brousseau, P., Malardel, S., Hello, G., Bénard, P., Bouttier, F., Lac, C., Masson, V.: The AROME-france convective-scale operational model. Mon. Weather Rev. **139**(3), 976–991 (2011). https://doi.org/10.1175/2010mwr3425.1
9. Wilks, D.S.: Statistical Methods in the Atmospheric Sciences, vol. 100, 3rd edn. Academic, Oxford (2011). https://www.sciencedirect.com/bookseries/international-geophysics/vol/100/suppl/C

One-Phase and Two-Phase Flow Simulation Using High-Order HDG and High-Order Diagonally Implicit Time Integration Schemes

Albert Costa-Solé, Eloi Ruiz-Gironés, and Josep Sarrate

Abstract We present two high-order hybridizable discontinuous Galerkin (HDG) formulations combined with high-order diagonally implicit Runge-Kutta schemes to solve one-phase and two-phase flow problems through porous media. The HDG method is locally conservative and allows reducing the size of the global systems due to the hybridization procedure, and the pressure, the saturation and the velocity converge with order $P + 1$ in L^2-norm, with P being the polynomial degree. In addition, an element-wise post-process can be applied to obtain a convergence rate of $P + 2$ in L^2-norm for the pressure and saturation. To achieve these rates of convergence the temporal errors should be small enough. For this purpose we combine HDG with high-order diagonally implicit Runge-Kutta (DIRK) temporal schemes. Finally, we present four examples dealing with 2D and 3D problems, and high-order structured and unstructured meshes.

1 Introduction

Mathematical models are the key stone in the management, planning, and analysis of oilfields exploitation. During the initial stages of hydrocarbon production (primary recovery), the pressure difference between the reservoir and the surface is high enough to move the hydrocarbons upward, see [7]. This stage approximately corresponds to 10% of the total oil production. One-phase flow through porous media is widely used to model this scenario. The governing equation is a non-linear transient partial differential equation (PDE), which is obtained from the combination

---

A. Costa-Solé · J. Sarrate (✉)
Laboratori de Càlcul Numèric (LaCàN), Universitat Politècnica de Catalunya, Barcelona, Spain
e-mail: albert.costa@upc.edu; jose.sarrate@upc.edu

E. Ruiz-Gironés
Barcelona Supercomputing Center – BSC, Barcelona, Spain
e-mail: eloi.ruizgirones@bsc.es

M. I. Asensio et al. (eds.), *Applied Mathematics for Environmental Problems*,
SEMA SIMAI Springer Series 6, https://doi.org/10.1007/978-3-030-61795-0_4

of the mass conservation with Darcy's law and equations of state for the fluid and the rock [4]. Once the pressure of the reservoir drops, a fluid (usually water) is injected to maintain the flow rate. This stage is known as secondary recovery and approximately accounts for 25% of the total oil production [7]. If a single hydrocarbon is considered and the pressure is above the bubble point, the two-phase immiscible flow model is widely used in industry to simulate this process [4]. The bubble point defines the limit that determines the hydrocarbon phases in the reservoir, see [4]. That is, above the bubble point the hydrocarbon phase is liquid, known as oil phase. Below the bubble point the hydrocarbon phase is liquid and gaseous. Thus, we are assuming that the hydrocarbon is only a fluid.

Numerical methods that accurately approximate the non-linear behavior of these models and properly capture the geometrical and heterogeneous complexity of hydrocarbon reservoirs are extremely demanding. Several techniques have been proposed to discretize in space and time both models. In this work, we propose to discretize these models using a high-order hybridizable discontinuous Galerkin (HDG) formulation in space combined with high-order diagonally implicit Runge-Kutta scheme in time.

On the one hand, HDG method exhibits several properties that make it attractive for these simulations, see [6, 9]. First, it allows using high-order unstructured meshes to appropriately discretize the reservoir. Second, it is conservative at the elemental level. Third, if the error of the temporal discretization is low enough, the scalar unknowns and the fluxes converge with $P + 1$ in the L^2-norm, with P being the polynomial degree [2]. Finally, an element-wise post-process can be applied at the desired time step to achieve a convergence rate of $P + 2$ in L^2-norm for the scalar variables, see [15, 18, 19, 23]. This allows us using large elements and in this way we are able to reduce the number of unknowns without penalizing the accuracy of the numerical solution. On the other hand, high-order diagonally implicit Runge-Kutta schemes allow using large time steps without hampering the accuracy [3, 14].

The combination of the HDG formulation in space with diagonally implicit Runge-Kutta schemes leads to an algebraic non-linear problem at each time stage of the temporal discretization.

For one-phase flow problems, we use a Newton-Raphson method to solve this non-linear problem. For two-phase flow problems, it is standard to solve the non-linear problem by splitting it into a pressure and a saturation equation. This approach reduces the memory requirements since there is no need to solve both variables at the same time. For instance, the IMPES method solves the pressure implicitly and the saturation explicitly [4]. Recently, there are DG formulations that apply this idea both using an explicit temporal scheme [1, 12, 13] and implicit temporal scheme [8, 9, 16]. To reduce the memory footprint of the global non-linear problem, we propose a fix point iterative procedure to alternatively solve implicitly the saturation and pressure equations.

The rest of the paper is organized as follows. Section 2 introduces the mathematical model for one-phase flow problems and details its numerical approximation when the HDG method combined with diagonally implicit Runge-Kutta schemes are used. Section 3 performs the same analysis for two-phase flow problems. Section 4 presents several examples including homogeneous and heterogeneous media of 2D and 3D domains discretized with structured and unstructured meshes. Finally, Sect. 5 summarizes the conclusions of the paper and briefly describes the future work.

2 Numerical Model for One-Phase Flow

In this section, we first state the mathematical model for one-phase flow through porous media problems. Then, we rewrite it as system of two first-order equations and deduce the corresponding HDG spatial discretization. This semi-discrete problem is a non-linear coupled system of first-order differential algebraic equations (DAEs) that is integrated using a diagonally implicit Runge-Kutta scheme. This leads to a non-linear system that is solved at each stage of the Runge-Kutta scheme using the Newton-Raphson method. Finally, we detail a local post-processing procedure that allows increasing the converge order of the obtained solution. We highlight that in our work we deal with temporal, diffusion, and convective non-linear terms, whereas in [19] only the convective term is non-linear.

2.1 Continuous Model

Let $\Omega \subset \mathbb{R}^d$ be a porous medium domain with boundary Γ such that $\Gamma = \Gamma_D \cup \Gamma_N$ and $\Gamma_D \cap \Gamma_N = \emptyset$, where Γ_D is the Dirichlet boundary and Γ_N is the Neumann boundary, and $T = (0, t_{\text{end}})$ a time interval. The governing equation for a single slightly compressible fluid under isothermal conditions that completely saturates the slightly compressible porous medium is obtained by combining the mass conservation equation and the Darcy's law, see details in [4]:

$$\left.\begin{aligned} \frac{\partial(\phi\rho)}{\partial t} + \nabla\cdot\left(-\frac{\rho}{\mu}\mathbf{K}\left(\nabla p - \rho\mathbf{g}\right)\right) &= f(\mathbf{x},t) && \forall(\mathbf{x},t)\in(\Omega,T)\\ p(\mathbf{x},t) &= g_D(\mathbf{x},t) && \forall(\mathbf{x},t)\in(\Gamma_D,T)\\ -\frac{\rho}{\mu}\mathbf{K}\left(\nabla p - \rho\mathbf{g}\right)\cdot\mathbf{n} &= g_N(\mathbf{x},t) && \forall(\mathbf{x},t)\in(\Gamma_N,T)\\ p(\mathbf{x},0) &= p^0(\mathbf{x}) && \forall\mathbf{x}\in\Omega \end{aligned}\right\} \tag{1}$$

where $\phi = \phi(p)$ is the porosity, $\rho = \rho(p)$ is the fluid density, μ is the fluid viscosity, t is the time, $\mathbf{K} = \text{diag}(\kappa_{11}, \kappa_{22}, \kappa_{33})$ is the soil absolute permeability tensor, $\mathbf{g}$ is the gravity, $f(\mathbf{x}, t)$ is the source term, $g_D(\mathbf{x}, t)$ and $g_N(\mathbf{x}, t)$ are the Dirichlet and Neumann prescribed values respectively, $\mathbf{n}$ is the outward normal, and $p^0(\mathbf{x})$ is the initial pressure of the reservoir.

We model the slightly compressible fluid and rock as

$$\rho(p) = \rho_{\text{ref}}\left(1 + c_f(p - p_{\text{ref}})\right), \qquad \phi(p) = \phi_{\text{ref}}\left(1 + c_r(p - p_{\text{ref}})\right),$$

where c_f and c_r are the fluid and rock compressibility, respectively, and ρ_{ref} and ϕ_{ref} are the reference density and the reference porosity at a reference pressure p_{ref}. In this work, we assume constant values for c_f and c_r, see [4].

In order to introduce the HDG formulation for one-phase flow, we rewrite Equation (1) as a system of two first-order equations. To this end, we introduce the notation

$$s(p) = \phi(p)\rho(p)c_t, \qquad \mathbf{A}(p) = \frac{\rho(p)}{\mu}\mathbf{K}, \qquad \mathbf{F}(p) = \frac{\rho(p)^2}{\mu}\mathbf{Kg}, \tag{2}$$

where $c_t = c_r + c_f$ is the total compressibility. Therefore, we obtain

$$\left.\begin{aligned} s(p)\frac{\partial p}{\partial t} + \nabla\cdot(\mathbf{q} + \mathbf{F}(p)) = f(\mathbf{x}, t) &\quad \forall(\mathbf{x}, t) \in (\Omega, T)\\ \mathbf{q} + \mathbf{A}(p)\nabla p = 0 &\quad \forall(\mathbf{x}, t) \in (\Omega, T)\\ p(\mathbf{x}, t) = g_D(\mathbf{x}, t) &\quad \forall(\mathbf{x}, t) \in (\Gamma_D, T)\\ (\mathbf{q} + \mathbf{F}(p))\cdot\mathbf{n} = g_N(\mathbf{x}, t) &\quad \forall(\mathbf{x}, t) \in (\Gamma_N, T)\\ p(\mathbf{x}, 0) = p^0(\mathbf{x}) &\quad \forall\mathbf{x} \in \Omega \end{aligned}\right\} \tag{3}$$

We identify $\mathbf{q} = -\mathbf{A}(p)\nabla p$ as the diffusive flux and $\mathbf{F}(p)$ as the convective flux.

2.2 HDG Spatial Discretization

We discretize the domain, Ω, with a tessellation, T_h, composed of a set of elements, e, of polynomial degree P. Afterwards, we introduce the discontinuous finite element spaces associated with the tessellation, T_h:

$$\mathbb{V}_h^P = \left\{v \in L^2\left(\Omega^d\right) \mid v_{|e} \in \left(\mathbb{S}^P(e)\right) \forall e \in \mathrm{T}_h\right\},$$

$$\mathbb{W}_h^P = \left\{\mathbf{w} \in \left(L^2\left(\Omega^d\right)\right)^d \mid \mathbf{w}_{|e} \in \left(\mathbb{S}^P(e)\right)^d \forall e \in \mathrm{T}_h\right\},$$

$$\mathbb{M}_h^P = \left\{\psi \in L^2(\Sigma_h) \mid \psi_{|f} \in \left(\mathbb{S}^P(f)\right) \forall f \in \Sigma_h\right\},$$

where $\mathbb{S}^P$ is the space of the polynomials of degree at most P for triangles and tetrahedra (usually denoted by $\mathbb{P}^P$), or the tensor products of polynomials of degree at most P in each coordinate direction for tensor product elements (usually denoted by $\mathbb{Q}^P$), d is the space dimension and Σ_h is the skeleton of the mesh, that is the set of all the element faces, f. We define $\mathbb{M}_h^P(g_D) = \{\psi \in \mathbb{M}_h^P \mid \psi = \Pi(g_D) \text{ on } \Gamma_D\}$, where $\Pi(\cdot)$ is a projection operator to the space $\{\psi|_{\Gamma_D} \forall \psi \in \mathbb{M}_h^P\}$. In this work, we use a fixed polynomial degree for all the elements. We also define the scalar products:

$$(a, b)_e = \int_e a\, b \,\mathrm{d}\Omega \quad \forall a, b \in \mathbb{V}_h^P,$$

$$(\mathbf{a}, \mathbf{b})_e = \int_e \mathbf{a} \cdot \mathbf{b} \,\mathrm{d}\Omega \quad \forall \mathbf{a}, \mathbf{b} \in \mathbb{W}_h^P,$$

$$\langle \hat{a}, \hat{b} \rangle_{\partial e} = \int_{\partial e} \hat{a}\, \hat{b} \,\mathrm{d}\Gamma \quad \forall \hat{a}, \hat{b} \in \mathbb{M}_h^P.$$

The HDG formulation seeks an approximation $(p_h, \mathbf{q}_h, \hat{p}_h) \in \mathbb{V}_h^P \times \mathbb{W}_h^P \times \mathbb{M}_h^P(g_D)$ such that

$$\left.\begin{aligned}
\sum_{e \in \mathrm{T}_h} \left(\left(s\,(p_h)\,\frac{\partial p_h}{\partial t}, v\right)_e - \left(\mathbf{q}_h + \mathbf{F}\,(p_h)\,, \nabla v\right)_e + \left\langle (\hat{\mathbf{q}}_h + \hat{\mathbf{F}}_h) \cdot \mathbf{n},\ v \right\rangle_{\partial e} \right) - & \\
\sum_{e \in \mathrm{T}_h} (f, v)_e &= 0 \\
\sum_{e \in \mathrm{T}_h} \left(\left(\mathbf{A}^{-1}\,(p_h)\,\mathbf{q}_h, \mathbf{w}\right)_e - (p_h, \nabla \cdot \mathbf{w})_e + \langle \hat{p}_h, \mathbf{w} \cdot \mathbf{n} \rangle_{\partial e} \right) &= 0 \\
\sum_{e \in \mathrm{T}_h} \left(\left\langle (\hat{\mathbf{q}}_h + \hat{\mathbf{F}}_h) \cdot \mathbf{n}, \psi \right\rangle_{\partial e} \right) - \langle g_N, \psi \rangle_{\Gamma_N} &= 0
\end{aligned}\right\} \tag{4}$$

for all $(v, \mathbf{w}, \psi) \in \mathbb{V}_h^P \times \mathbb{W}_h^P \times \mathbb{M}_h^P(0)$, where $\hat{p}_h$ is the trace of the pressure defined on the mesh skeleton, Σ_h, and $\hat{\mathbf{q}}_h + \hat{\mathbf{F}}_h$ is the total numerical flux. The third equation in (4) is the transmissivity equation, in which we impose the continuity of the total numerical flux in the normal direction between adjacent elements, see [15]. Therefore, this equation relates the unknowns between adjacent elements.

According to [18, 19], we define the total numerical flux as

$$\hat{\mathbf{q}}_h + \hat{\mathbf{F}}_h = \mathbf{q}_h + \mathbf{F}\left(\hat{p}_h\right) + \tau(p_h, \hat{p}_h)(p_h - \hat{p}_h)\,\mathbf{n}, \qquad \text{on } \Sigma_h,$$

where τ is the stabilization parameter that depends on p_h and $\hat{p}_h$. Nevertheless, to facilitate the notation, from now on, we will not write explicitly this dependency.

Following [18, 19], we split the stabilization parameter into a diffusive and convective terms as

$$\tau = \tau_{\text{diff}} + \tau_{\text{conv}},$$

and we set the diffusive and convective numerical fluxes on Σ_h as

$$\hat{\mathbf{q}}_h = \mathbf{q}_h + \tau_{\text{diff}}(p_h - \hat{p}_h)\,\mathbf{n}, \qquad \hat{\mathbf{F}}_h = \mathbf{F}\left(\hat{p}_h\right) + \tau_{\text{conv}}(p_h - \hat{p}_h)\,\mathbf{n}, \tag{5}$$

respectively. We define the diffusive stabilization parameter as

$$\tau_{\text{diff}} = \frac{1}{l_c}\frac{\rho(p_h)}{\mu}\gamma_{\mathbf{K}}, \tag{6}$$

where l_c is a characteristic length of the problem, and $\gamma_{\mathbf{K}}$ the maximum eigenvalue of the permeability tensor, $\mathbf{K}$.

To select the τ_{conv} we use a monotone scheme flux, which ensures the stability of the numerical method [18, 19]. Specifically, we define τ_{conv} as

$$\tau_{\text{conv}} = \frac{1}{(p_h - \hat{p}_h)^2}\int_{\hat{p}_h}^{p_h}\left(\frac{\widehat{\mathbf{F}\cdot\mathbf{n}}^{EO}(s, \hat{p}_h) - \mathbf{F}(\hat{p}_h)\cdot\mathbf{n}}{p_h - \hat{p}_h}\right)\mathrm{d}s, \tag{7}$$

where $\widehat{\mathbf{F}\cdot\mathbf{n}}^{EO}(\cdot,\cdot)$ is the Engquist-Osher monotone scheme flux that for this problem can be computed analytically.

We highlight that in this work all the Dirichlet boundary conditions are applied using the traces of the pressure as follows:

$$\hat{p}_h = \Pi(g_D) \qquad \forall \mathbf{x} \in \partial \mathrm{T}_{h\Gamma_D},$$

where $\partial \mathrm{T}_{h\Gamma_D}$ is the set of mesh faces on the Dirichlet boundary.

Let $\{N_i\}_{i=1,\ldots,N}$ be a Lagrangian basis of shape functions of $\mathbb{S}^P(e)$, where N is the total number of element nodes, and let $\{N_l^f\}_{l=1,\ldots,N_f}$ be a Lagrangian basis of shape functions of $\mathbb{S}^P(f)$, where N_f is the total number of nodes on the element faces. We define the approximations p_h, $\mathbf{q}_h$, $\hat{p}_h$, and $\dot{p}_h = \partial p_h/\partial t$ as

$$\begin{aligned} p_h(\mathbf{x}, t) &= \sum_{e\in \mathrm{T}_h}\sum_{i=1}^{N} p_i(t)N_i(\mathbf{x}) \qquad \mathbf{q}_h(\mathbf{x}, t) = \sum_{e\in \mathrm{T}_h}\sum_{i=1}^{N}\sum_{j=1}^{N_{sd}} q_{i,j}(t)N_i(\mathbf{x})\mathbf{e}_j \\ \hat{p}_h(\mathbf{x}, t) &= \sum_{f\in \Sigma_h}\sum_{l=1}^{N_f} \hat{p}_l(t)N_l^f(\mathbf{x}) \qquad \dot{p}_h(\mathbf{x}, t) = \sum_{e\in \mathrm{T}_h}\sum_{i=1}^{N} \dot{p}_i(t)N_i(\mathbf{x}) \end{aligned} \tag{8}$$

where N_{sd} is the physical dimension of the problem, and $\dot{p}_i(t) = dp_i(t)/dt$.

Inserting the diffusive and convective fluxes (5) and the approximations (8) into Equations (4) we obtain a non-linear coupled system of first-order DAE that can be written as

$$\mathbf{R}\left(t, \mathbf{p}, \dot{\mathbf{p}}, \mathbf{q}, \hat{\mathbf{p}}\right) = \begin{bmatrix} \mathbf{R}_{\dot{p}}\left(t, \mathbf{p}, \dot{\mathbf{p}}, \mathbf{q}, \hat{\mathbf{p}}\right) \\ \mathbf{R}_{\mathbf{q}}\left(t, \mathbf{p}, \dot{\mathbf{p}}, \mathbf{q}, \hat{\mathbf{p}}\right) \\ \mathbf{R}_{\hat{p}}\left(t, \mathbf{p}, \dot{\mathbf{p}}, \mathbf{q}, \hat{\mathbf{p}}\right) \end{bmatrix} = \mathbf{0}, \tag{9}$$

where $\mathbf{p}, \dot{\mathbf{p}}, \mathbf{q}, \hat{\mathbf{p}}$ are vectors composed of all the nodal values for the pressure, $p_i(t)$, the pressure derivative, $\dot{p}_i(t)$, the numerical flux, $q_{i,j}(t)$, and the trace of the pressure, $\hat{p}_l(t)$ at time t.

Thus, given an approximation of $(p_h, \dot{p}_h, \mathbf{q}_h, \hat{p}_h) \in \mathbb{V}_h^P \times \mathbb{V}_h^P \times \mathbb{W}_h^P \times \mathbb{M}_h^P(g_D)$, $\mathbf{R}_{\dot{p}}$, $\mathbf{R}_{\mathbf{q}}$ and $\mathbf{R}_{\hat{p}}$ are defined as follows:

$$\begin{aligned}
\left[\mathbf{R}_{\dot{p}}\right]_i &= \sum_{e \in \mathrm{T}_h} \left((s\dot{p}_h, N_i)_e - (\mathbf{q}_h + \mathbf{F}_h, \nabla N_i)_e + \langle \hat{\mathbf{F}}_h \cdot \mathbf{n},\ N_i \rangle_{\partial e} \right) \\
&\quad + \sum_{e \in \mathrm{T}_h} \left(\langle \mathbf{q}_h \cdot \mathbf{n} + \tau_{\text{diff}}(p_h - \hat{p}_h),\ N_i \rangle_{\partial e} - (f, N_i)_e \right), \\
\left[\mathbf{R}_{\mathbf{q}}\right]_{i,j} &= \sum_{e} \left((\mathbf{A}^{-1}\mathbf{q}_h, N_i \mathbf{e}_j)_e - \left(p_h, \nabla \cdot (N_i \mathbf{e}_j)\right)_e + \langle \hat{p}_h, N_i \mathbf{e}_j \cdot \mathbf{n} \rangle_{\partial e} \right), \\
\left[\mathbf{R}_{\hat{p}}\right]_l &= \sum_{e \in \mathrm{T}_h} \left(\langle \hat{\mathbf{F}}_h \cdot \mathbf{n},\ N_l^f \rangle_{\partial e} + \langle \mathbf{q}_h \cdot \mathbf{n} + \tau_{\text{diff}}(p_h - \hat{p}_h),\ N_l^f \rangle_{\partial e} \right) - \langle g_N, N_l^f \rangle_{\Gamma_N}.
\end{aligned}$$

2.3 *Temporal Discretization*

To integrate in time the DAE in Equation (9), we use a diagonally implicit Runge-Kutta method (DIRK). From now on, we denote by $(\cdot)^n$ the value of any variable at time t^n, and by $(\cdot)^{n,i}$ the value of any variable at time $t^{n,i} = t^n + c_i \Delta t$, with n being the time step and i the DIRK stage. Accordingly, we compute the pressure at time $t^{n+1} = t^n + \Delta t$ as

$$\mathbf{p}^{n+1} = \mathbf{p}^n + \Delta t \sum_{i=1}^{s} b_i \dot{\mathbf{p}}^{n,i},$$

where s is the number of stages and $\dot{\mathbf{p}}^{n,i}$ is the approximation of $\dot{\mathbf{p}}$ at time $t^{n,i}$. The pressure at each stage of the DIRK scheme is computed as

$$\mathbf{p}^{n,i} = \mathbf{p}^n + \Delta t \sum_{j=1}^{i} a_{ij} \dot{\mathbf{p}}^{n,j},$$

Table 1 Butcher's table for a diagonal implicit Runge-Kutta scheme

c_1	a_{11}			
c_2	a_{21}	a_{22}		
$\vdots$	$\vdots$		$\ddots$	
c_s	a_{s1}		$\dots$	a_{ss}
	b_1	b_2	$\dots$	b_s

and the $\dot{\mathbf{p}}^{n,i}$ for $i = 1, \dots, s$ is computed as the solution of the non-linear algebraic equation

$$\mathbf{R}\left(t^{n,i}, \mathbf{q}^{n,i}, \mathbf{p}^n + \Delta t \sum_{j=1}^{i} a_{ij}\dot{\mathbf{p}}^{n,j}, \dot{\mathbf{p}}^{n,i}, \hat{\mathbf{p}}^{n,i}\right) = \mathbf{0}. \tag{10}$$

The parameters b_i, c_i, a_{ij}, with $i = 1, \dots, s$ and $j = 1, \dots, i$, define the DIRK method and are given by Butcher's tables, see Table 1, [3, 14, 17].

2.4 *Non-linear Solver*

To solve the non-linear system (10) we use the Newton-Raphson method. Hence, we define the global unknown

$$\mathbf{u}^{n,i} = \begin{bmatrix} \dot{\mathbf{p}}^{n,i} \\ \mathbf{q}^{n,i} \\ \hat{\mathbf{p}}^{n,i} \end{bmatrix},$$

and at each iteration we solve the linear system

$$\mathbf{J}\left(\mathbf{u}^{n,i,k}\right)\delta\mathbf{u}^{n,i,k} = -\mathbf{R}\left(\mathbf{u}^{n,i,k}\right),$$

where $\mathbf{u}^{n,i,k}$ is the k-th approximation of $\mathbf{u}^n$ at i-th Runge-Kutta stage, $\delta\mathbf{u}^{n,i,k} = \mathbf{u}^{n,i,k+1} - \mathbf{u}^{n,i,k}$, and $\mathbf{J}\left(\mathbf{u}^{n,i,k}\right)$ is the Jacobian matrix of $\mathbf{R}$ evaluated at $\mathbf{u}^{n,i,k}$.

We define the stopping criteria of the non-linear solver as

$$\begin{aligned} &\frac{\|p_h^{n,i,k} - p_h^{n,i,k+1}\|_{L^2(\mathrm{T}_h)}}{\|p_h^{n,i,k+1}\|_{L^2(\mathrm{T}_h)}} \le \varepsilon_p, \qquad \frac{\|\mathbf{q}_h^{n,i,k} - \mathbf{q}_h^{n,i,k+1}\|_{L^2(\mathrm{T}_h)}}{\|\mathbf{q}_h^{n,i,k+1}\|_{L^2(\mathrm{T}_h)}} \le \varepsilon_{\mathbf{q}}, \\ &\frac{\|\hat{p}_h^{n,i,k} - \hat{p}_h^{n,i,k+1}\|_{L^2(\Sigma_h)}}{\|\hat{p}_h^{n,i,k+1}\|_{L^2(\Sigma_h)}} \le \varepsilon_{\hat{p}}, \qquad \|\mathbf{R}\left(\mathbf{u}^{n,i,k}\right)\|_2 \le \varepsilon_R. \end{aligned} \tag{11}$$

where ε_p, $\varepsilon_{\mathbf{q}}$, $\varepsilon_{\hat{p}}$, and ε_R are four prescribed tolerances; $\|\cdot\|_2$ is the Euclidean norm of vectors; and $\|\cdot\|_{L^2(\mathrm{T}_h)}$ and $\|\cdot\|_{L^2(\Sigma_h)}$ are the L^2 norm of functions on T_h and Σ_h respectively.

2.5 *Local Post-processing*

One of the main advantages of using the HDG formulation is that both the pressure, p_h, and its flux, $\mathbf{q}_h$, in $\mathbb{V}_h^P$ and $\mathbb{W}_h^P$ spaces, respectively, have a rate of convergence of $P+1$ in the L^2-norm, when the temporal error is low enough. Moreover, a local post-processing can be applied to obtain a new approximation for the pressure, p_h^*, in $\mathbb{V}_h^{P+1}$ with a rate of convergence of $P+2$ in the L^2-norm, see [15, 18, 19].

The local problem consists of finding the post-processed pressure, $p_h^* \in \mathbb{V}_h^{P+1}$, on each element, e, such that

$$\left.\begin{aligned} (\mathbf{A}(p_h)\nabla p_h^*, \nabla v)_e &= -(\mathbf{q}_h, \nabla v)_e \\ (p_h^*, 1)_e &= (p_h, 1)_e \end{aligned}\right\} \tag{12}$$

for all $v \in \mathbb{V}_h^{P+1}$.

In order to obtain a well-posed and invertible system, the second equation in (12) is added, which imposes that the averages of the post-processed pressure, p_h^*, and the approximated pressure, p_h, are equal element by element. According to [15, 18, 19] it is important to highlight that this procedure can be applied at selected time steps, and it is not necessary to apply it to all the time steps.

3 Numerical Model for Two-Phase Flow

In this section, we pose the mathematical model for two-phase flow through porous media. We select the water pressure and the oil saturation as main unknowns, and rewrite the model as a system of four first-order equations. Similarly to one-phase problem, we detail the corresponding HDG spatial discretization, the diagonally implicit Runge-Kutta scheme for time integration, and the non-linear solver. We use a fix-point procedure to solve the non-linear system at each stage of the DIRK scheme. Specifically, we alternatively solve a non-linear problem for the saturation unknowns, and then we solve a linear problem for the pressure unknowns implicitly until convergence is achieved. The proposed method differs from the classical IMPES method, since the latter solves explicitly the saturation and implicitly the pressure.

3.1 Continuous Model

The governing equations for two-phase flow through porous media are provided by the mass conservation and the Darcy's law for each phase [2, 4]:

$$\frac{\partial(\phi\rho_\alpha S_\alpha)}{\partial t} + \nabla\cdot(\rho_\alpha \mathbf{v}_\alpha) = \rho_\alpha f_\alpha \qquad \alpha = w, o,$$

$$\mathbf{v}_\alpha = -\lambda_\alpha \mathbf{K}\nabla p_\alpha \qquad \alpha = w, o,$$

where w stands for the wetting phase (water), o stands for the non-wetting phase (oil), S_α are the saturation for the water and oil, respectively, and $\lambda_\alpha = \kappa_{r\alpha}/\mu_\alpha$ is the phase mobility, with $\kappa_{r\alpha}$ and μ_α being the relative permeability and the viscosity of phase α, respectively.

We assume that both phases fill the voids of the soil, and that there is a discontinuity in the pressure field due to the interface tension between phases called capillary pressure, p_c, see details in [2, 4]. That is:

$$S_w + S_o = 1, \qquad p_c = p_o - p_w.$$

Under these assumptions, the capillary pressure, p_c, and the relative permeabilities of each phase, $\kappa_{r\alpha}$, are related to the water or oil saturations. In this work we use the Brooks-Corey model, see [5]:

$$\left.\begin{aligned} p_c &= p_e(1-S_{eo})^{-1/\theta} \\ \kappa_{rw} &= (1-S_{eo})^{(2+3\theta)/\theta} \\ \kappa_{ro} &= {S_{eo}}^2\left(1-(1-S_{eo})^{(2+\theta)/\theta}\right) \end{aligned}\right\} \tag{13}$$

where p_e is the entry pressure, θ is the pore size distribution, and

$$S_{eo} = \frac{S_o - S_{ro}}{1 - S_{rw} - S_{ro}}$$

is the effective oil saturation, with S_{ro} and S_{rw} being the residual oil and water saturation, respectively.

We consider a domain Ω and time interval $T = (0, t_{\text{end}})$. The boundary of Ω is divided in three disjoint parts such that $\partial\Omega = \Gamma_{in} \cup \Gamma_{\text{out}} \cup \Gamma_{nf}$, where Γ_{in} is the inflow boundary (water injection), Γ_{out} is the outflow boundary (water and oil extraction), and Γ_{nf} is the no-flow boundary. Assuming immiscible and incompressible fluids, incompressible rock, and selecting the water pressure, p_w, and the oil saturation, S_o, as main unknowns [2, 4], we obtain a system of two coupled non-linear equations. On the one hand, the water pressure equation is

$$\left.\begin{aligned} -\nabla\cdot(\lambda_t\mathbf{K}\nabla p_w+\lambda_o\mathbf{K}\nabla p_c) &= f_o+f_w && \forall(\mathbf{x},t)\in(\Omega,T)\\ p_w{}^{\Gamma_{\text{in}}} &= g_{Dp}^{\text{in}} && \forall(\mathbf{x},t)\in(\Gamma_{\text{in}},T)\\ p_w{}^{\Gamma_{\text{out}}} &= g_{Dp}^{\text{out}} && \forall(\mathbf{x},t)\in(\Gamma_{\text{out}},T)\\ \mathbf{v}_t\cdot\mathbf{n} &= 0 && \forall(\mathbf{x},t)\in(\Gamma_{\text{nf}},T) \end{aligned}\right\} \quad (14)$$

where $\lambda_t=\lambda_o+\lambda_w$ is the total mobility, $\mathbf{v}_t=\mathbf{v}_o+\mathbf{v}_w$ is the total velocity, and $g_{Dp}^{in}, g_{Dp}^{\text{out}}$ are the values of the Dirichlet boundaries condition for the pressure on the inflow and outflow boundaries, respectively.

On the other hand, the oil saturation equation is

$$\left.\begin{aligned} \phi\frac{\partial S_o}{\partial t}-\nabla\cdot(\lambda_o\mathbf{K}(\nabla p_c+\nabla p_w)) &= f_o && \forall(\mathbf{x},t)\in(\Omega,T)\\ S_o{}^{\Gamma_{\text{in}}} &= g_{Ds}^{\text{in}} && \forall(\mathbf{x},t)\in(\Gamma_{\text{in}},T)\\ \left(\frac{\lambda_o\lambda_w}{\lambda_t}\mathbf{K}\nabla p_c\right)\cdot\mathbf{n} &= g_{Ns}^{\text{out}} && \forall(\mathbf{x},t)\in(\Gamma_{\text{out}},T)\\ \mathbf{v}_o\cdot\mathbf{n} &= 0 && \forall(\mathbf{x},t)\in(\Gamma_{\text{nf}},T)\\ S_o(\cdot,0) &= S_o^0(\mathbf{x}) && \forall\mathbf{x}\in\Omega \end{aligned}\right\} \quad (15)$$

where $g_{Ds}^{in}, g_{Ds}^{\text{out}}$ are the Dirichlet boundary condition values of the saturation for the inflow and outflow boundary respectively, and g_{Ns}^{out} is the value of the Neumann boundary condition at the output boundaries.

In order to introduce the HDG formulation for two-phase flow, we rewrite equations (14) and (15) as a system of first-order PDEs by using the diffusive fluxes:

$$\mathbf{q}_p=-\lambda_t\mathbf{K}\nabla p_w, \qquad \mathbf{q}_s=-\lambda_o\mathbf{K}\nabla p_c,$$

see details in [15, 18, 19]. Specifically, the water pressure equation becomes

$$\left.\begin{aligned} \nabla\cdot(\mathbf{q}_p+\mathbf{q}_s) &= f_o+f_w && \forall(\mathbf{x},t)\in(\Omega,T)\\ \mathbf{q}_p+\lambda_t\mathbf{K}\nabla p_w &= \mathbf{0} && \forall(\mathbf{x},t)\in(\Omega,T)\\ p_w{}^{\Gamma_{\text{in}}} &= g_{Dp}^{\text{in}} && \forall(\mathbf{x},t)\in(\Gamma_{\text{in}},T)\\ p_w{}^{\Gamma_{\text{out}}} &= g_{Dp}^{\text{out}} && \forall(\mathbf{x},t)\in(\Gamma_{\text{out}},T)\\ \mathbf{v}_t\cdot\mathbf{n} &= 0 && \forall(\mathbf{x},t)\in(\Gamma_{\text{nf}},T) \end{aligned}\right\} \quad (16)$$

Similarly, the oil saturation equation becomes

$$\left.\begin{aligned} \phi\frac{\partial S_o}{\partial t} + \nabla\cdot\left(\mathbf{q}_s + \frac{\lambda_o}{\lambda_t}\mathbf{q}_p\right) &= f_o && \forall(\mathbf{x},t)\in(\Omega,T)\\ \mathbf{q}_s + \lambda_o\mathbf{K}\nabla p_c &= \mathbf{0} && \forall(\mathbf{x},t)\in(\Omega,T)\\ S_o{}^{\Gamma_{\text{in}}} &= g_{Ds}^{\text{in}} && \forall(\mathbf{x},t)\in(\Gamma_{\text{in}},T)\\ \left(\frac{\lambda_o\lambda_w}{\lambda_t}\mathbf{K}\nabla p_c\right)\cdot\mathbf{n} &= g_{Ns}^{\text{out}} && \forall(\mathbf{x},t)\in(\Gamma_{\text{out}},T)\\ \mathbf{v}_o\cdot\mathbf{n} &= 0 && \forall(\mathbf{x},t)\in(\Gamma_{\text{nf}},T)\\ S_o(\cdot,0) &= S_o^0(\mathbf{x}) && \forall\mathbf{x}\in\Omega \end{aligned}\right\} \tag{17}$$

3.2 *HDG Spatial Discretization*

We discretize the domain, Ω, with a tessellation, T_h, composed of a set of elements, e, of polynomial degree P, and we use the same discontinuous spaces and the scalar products introduced in Sect. 2.2.

The HDG formulation for the water pressure corresponding to Equation (16) seeks an approximation $(p_{w_h}, \mathbf{q}_{p_h}, \hat{p}_{w_h}) \in \mathbb{V}_h^P \times \mathbb{W}_h^P \times \mathbb{M}_h^P(g_D)$ such that:

$$\left.\begin{aligned} \sum_{e\in\mathrm{T}_h}\left(-(\mathbf{q}_{p_h}+\mathbf{q}_{s_h},\nabla v)_e + \langle(\hat{\mathbf{q}}_{p_h}+\hat{\mathbf{q}}_{s_h})\cdot\mathbf{n}, v\rangle_{\partial e}\right) &= \\ \sum_{e\in\mathrm{T}_h}((f_o+f_w, v)_e)& \\ \sum_{e\in\mathrm{T}_h}\left((\mathbf{A}_{p_h}^{-1}\mathbf{q}_{p_h},\mathbf{w})_e - (p_{w_h},\nabla\cdot\mathbf{w})_e + \langle\hat{p}_{w_h},\mathbf{w}\cdot\mathbf{n}\rangle_{\partial e}\right) &= 0\\ \sum_{e\in\mathrm{T}_h}\langle(\hat{\mathbf{q}}_{p_h}+\hat{\mathbf{q}}_{s_h})\cdot\mathbf{n},\psi\rangle_{\partial e} &= 0 \end{aligned}\right\} \tag{18}$$

for all $(v, \mathbf{w}, \psi) \in \mathbb{V}_h^P \times \mathbb{W}_h^P \times \mathbb{M}_h^P(0)$, where $\mathbf{A}_{p_h} = \lambda_t\mathbf{K}$.

The HDG formulation for the oil saturation corresponding to Equation (17) seeks an approximation $(S_{o_h}, \mathbf{q}_{s_h}, \hat{S}_{oh}) \in \mathbb{V}_h^P \times \mathbb{W}_h^P \times \mathbb{M}_h^P(g_D)$ such that:

$$\left.\begin{aligned}
&\sum_{e\in T_h}\left(\left(\phi\frac{\partial S_{o_h}}{\partial t}, v\right)_e - \left(\mathbf{q}_{s_h} + \frac{\lambda_o}{\lambda_t}\mathbf{q}_{p_h}, \nabla v\right)_e\right) + \\
&\qquad\sum_{e\in T_h}\left(\left\langle\left(\hat{\mathbf{q}}_{s_h} + \frac{\hat{\lambda}_o}{\hat{\lambda}_t}\hat{\mathbf{q}}_{p_h}\right)\cdot\mathbf{n}, v\right\rangle\right)_{\partial e} = \sum_{e\in T_h}(f_o, v)_e \\
&\sum_{e\in T_h}\left((\mathbf{A}_{s_h}^{-1}\mathbf{q}_{s_h})_e - (S_{o_h}, \nabla\cdot\mathbf{w})_e + \langle\hat{S}_{o_h}, \mathbf{w}\cdot\mathbf{n}\rangle_{\partial e}\right) = 0 \\
&\sum_{e\in T_h}\left(\left\langle\left(\hat{\mathbf{q}}_{s_h} + \frac{\hat{\lambda}_o}{\hat{\lambda}_t}\hat{\mathbf{q}}_{p_h}\right)\cdot\mathbf{n}, \psi\right\rangle_{\partial e}\right) - \langle g_{Ns}^{\text{out}}, \psi\rangle_{\Gamma_N^s} = 0
\end{aligned}\right\} \quad (19)$$

for all $(v, \mathbf{w}, \psi) \in \mathbb{V}_h^P \times \mathbb{W}_h^P \times \mathbb{M}_h^P(0)$, where $\hat{\lambda}_o$ and $\hat{\lambda}_t$ are the oil phase mobility and the total phase mobility evaluated using the traces, respectively, and $\mathbf{A}_{s_h} = \lambda_o p_c' \mathbf{K}$, with p_c' being the derivative of the capillary pressure respect to the oil saturation.

We define numerical flux for the water pressure and the numerical flux for the oil saturation as

$$\hat{\mathbf{q}}_{p_h} = \mathbf{q}_{p_h} + \tau_p(p_{w_h} - \hat{p}_{w_h})\,\mathbf{n}, \quad (20)$$

$$\hat{\mathbf{q}}_{s_h} = \mathbf{q}_{s_h} + \tau_s(S_{o_h} - \hat{S}_{o_h})\,\mathbf{n}, \quad (21)$$

where τ_p is a stabilization function for the water pressure and τ_s is a stabilization function for the oil saturation. According to [18, 19], we set the diffusive stabilization parameter of Equations (20) and (21) as

$$\tau_p = \frac{\hat{\lambda}_t}{l_p}\gamma_{\mathbf{K}}, \qquad \tau_s = \frac{\hat{\lambda}_o p_c'}{l_s}\gamma_{\mathbf{K}}, \quad (22)$$

respectively, where $\gamma_{\mathbf{K}}$ is the maximum eigenvalue of the permeability matrix, $\mathbf{K}$, l_p is the characteristic length for the pressure, and l_s is the characteristic length for the saturation.

Similarly to one-phase flow discretization, we define S_{o_h}, p_{w_h}, $\mathbf{q}_{s_h}$, $\mathbf{q}_{p_h}$, $\hat{S}_{oh}$, $\hat{p}_{w_h}$, and $\dot{S}_{o_h} = \partial S_{o_h}/\partial t$ as

$$
\begin{aligned}
S_{o_h}(\mathbf{x},t) &= \sum_{e\in \mathrm{T}_h}\sum_{i=1}^{N} S_i(t)N_i(\mathbf{x}) & \mathbf{q}_{s_h}(\mathbf{x},t) &= \sum_{e\in \mathrm{T}_h}\sum_{i=1}^{N}\sum_{j=1}^{N_{sd}} q_{s\,i,j}(t)N_i(\mathbf{x})\mathbf{e}_j \\
p_{w_h}(\mathbf{x},t) &= \sum_{e\in \mathrm{T}_h}\sum_{i=1}^{N} p_i(t)N_i(\mathbf{x}) & \mathbf{q}_{p_h}(\mathbf{x},t) &= \sum_{e\in \mathrm{T}_h}\sum_{i=1}^{N}\sum_{j=1}^{N_{sd}} q_{p_{i,j}}(t)N_i(\mathbf{x})\mathbf{e}_j \\
\hat{S}_{o_h}(\mathbf{x},t) &= \sum_{f\in \Sigma_h}\sum_{i=1}^{N_f} \hat{S}_i(t)N_l^f(\mathbf{x}) & \hat{p}_{w_h}(\mathbf{x},t) &= \sum_{f\in \Sigma_h}\sum_{i=1}^{N_f} \hat{p}_i(t)N_l^f(\mathbf{x}) \\
\dot{S}_{o_h}(\mathbf{x},t) &= \sum_{e\in \mathrm{T}_h}\sum_{i=1}^{N} \dot{S}_i(t)N_i(\mathbf{x})
\end{aligned}
\tag{23}
$$

where $\dot{S}_i(t) = dS_i(t)/dt$.

Inserting the numerical fluxes Equations (20) and (21) into Equations (18) and (19), and using the discretizations detailed in (23), we obtain a non-linear coupled system of first-order DAE

$$
\mathbf{R}\left(t,\mathbf{S}_o,\dot{\mathbf{S}}_o,\mathbf{q}_s,\hat{\mathbf{S}}_o,\mathbf{p}_w,\mathbf{q}_p,\hat{\mathbf{p}}_w\right) = \begin{bmatrix}
\mathbf{R}_p\left(t,\mathbf{S}_o,\dot{\mathbf{S}}_o,\mathbf{q}_s,\hat{\mathbf{S}}_o,\mathbf{p}_w,\mathbf{q}_p,\hat{\mathbf{p}}_w\right) \\
\mathbf{R}_{\mathbf{q}_p}\left(t,\mathbf{S}_o,\dot{\mathbf{S}}_o,\mathbf{q}_s,\hat{\mathbf{S}}_o,\mathbf{p}_w,\mathbf{q}_p,\hat{\mathbf{p}}_w\right) \\
\mathbf{R}_{\hat{p}}\left(t,\mathbf{S}_o,\dot{\mathbf{S}}_o,\mathbf{q}_s,\hat{\mathbf{S}}_o,\mathbf{p}_w,\mathbf{q}_p,\hat{\mathbf{p}}_w\right) \\
\mathbf{R}_{\dot{S}}\left(t,\mathbf{S}_o,\dot{\mathbf{S}}_o,\mathbf{q}_s,\hat{\mathbf{S}}_o,\mathbf{p}_w,\mathbf{q}_p,\hat{\mathbf{p}}_w\right) \\
\mathbf{R}_{\mathbf{q}_s}\left(t,\mathbf{S}_o,\dot{\mathbf{S}}_o,\mathbf{q}_s,\hat{\mathbf{S}}_o,\mathbf{p}_w,\mathbf{q}_p,\hat{\mathbf{p}}_w\right) \\
\mathbf{R}_{\hat{S}}\left(t,\mathbf{S}_o,\dot{\mathbf{S}}_o,\mathbf{q}_s,\hat{\mathbf{S}}_o,\mathbf{p}_w,\mathbf{q}_p,\hat{\mathbf{p}}_w\right)
\end{bmatrix} = \mathbf{0},
\tag{24}
$$

where $\mathbf{S}_o$, $\dot{\mathbf{S}}_o$, $\mathbf{q}_s$, $\hat{\mathbf{S}}_o$, $\mathbf{p}_w$, $\mathbf{q}_p$, $\hat{\mathbf{p}}_w$ are vectors composed of all the nodal values for the oil saturation, $S_i(t)$, the derivative of the oil saturation, $\dot{S}_i(t)$, the numerical flux for the oil saturation, $q_{s\,i,j}(t)$, the trace of the oil saturation, $\hat{S}_i(t)$, the water pressure, $p_i(t)$, the numerical flux for the water pressure, $q_{p_{i,j}}(t)$, and the traces of the water pressure, $\hat{p}_i(t)$.

Thus, given an approximation of $(p_{w_h}, \dot{p}_{w_h}, \mathbf{q}_{p_h}, \hat{p}_{w_h}) \in \mathbb{V}_h^P \times \mathbb{V}_h^P \times \mathbb{W}_h^P \times \mathbb{M}_h^P(g_D)$, and an approximation of $(S_{o_h}, \dot{S}_{o_h}, \mathbf{q}_{s_h}, \hat{p}_{w_h}) \in \mathbb{V}_h^P \times \mathbb{V}_h^P \times \mathbb{W}_h^P \times \mathbb{M}_h^P(g_D)$, the residuals $\mathbf{R}_p$, $\mathbf{R}_{\mathbf{q}_p}$, $\mathbf{R}_{\hat{p}}$, $\mathbf{R}_{\dot{S}}$, $\mathbf{R}_{\mathbf{q}_s}$, $\mathbf{R}_{\hat{S}}$ are defined as follows:

$$\begin{aligned}
\left[\mathbf{R}_p\right]_i &= \sum_{e\in \mathrm{T}_h} \left(-(\mathbf{q}_{p_h} + \mathbf{q}_{s_h}, \nabla N_i)_e + \langle \mathbf{q}_{p_h} \cdot \mathbf{n} + \tau_p(p_{w_h} - \hat{p}_{w_h}), N_i\rangle_{\partial e}\right) \\
&+ \sum_{e\in \mathrm{T}_h} \left(\langle \mathbf{q}_{s_h} \cdot \mathbf{n} + \tau_s(S_{o_h} - \hat{S}_{o_h}), N_i\rangle_{\partial e}\right) - \sum_{e\in \mathrm{T}_h} (f_o + f_w, N_i)_e \\
\left[\mathbf{R}_{\mathbf{q}_p}\right]_{i,j} &= \sum_{e\in \mathrm{T}_h} \left((\mathbf{A}_{p_h}^{-1}\mathbf{q}_{p_h}, N_i\mathbf{e}_j)_e - (p_{w_h}, \nabla\cdot(N_i\mathbf{e}_j))_e + \langle \hat{p}_{w_h}, N_i\mathbf{e}_j\cdot\mathbf{n}\rangle_{\partial e}\right) \\
\left[\mathbf{R}_{\hat{p}}\right]_l &= \sum_{e\in \mathrm{T}_h} \left(\langle \mathbf{q}_{p_h}\cdot\mathbf{n} + \tau_p(p_{w_h} - \hat{p}_{w_h}) + \mathbf{q}_{s_h}\cdot\mathbf{n} + \tau_s(S_{o_h} - \hat{S}_{o_h}), N_l^f\rangle_{\partial e}\right) \\
\left[\mathbf{R}_{\dot{S}}\right]_i &= \sum_{e\in \mathrm{T}_h} \left(\left(\phi\frac{\partial S_{o_h}}{\partial t}, N_i\right)_e - \left(\mathbf{q}_{s_h} + \frac{\lambda_o}{\lambda_t}\mathbf{q}_{p_h}, \nabla N_i\right)_e\right) \\
&+ \sum_{e\in \mathrm{T}_h} \left(\left\langle \mathbf{q}_{s_h}\cdot\mathbf{n} + \tau_s(S_{o_h} - \hat{S}_{o_h}), N_i\right\rangle_{\partial e}\right) \\
&+ \sum_{e\in \mathrm{T}_h} \left(\left\langle \frac{\hat{\lambda}_o}{\hat{\lambda}_t}\left(\mathbf{q}_{p_h}\cdot\mathbf{n} + \tau_p(p_{w_h} - \hat{p}_{w_h})\right), N_i\right\rangle_{\partial e}\right) - \sum_{e\in \mathrm{T}_h} (f_o, N_i)_e \\
\left[\mathbf{R}_{\mathbf{q}_s}\right]_{i,j} &= \sum_{e\in \mathrm{T}_h} \left((\mathbf{A}_{s_h}^{-1}\mathbf{q}_{s_h}, N_i\mathbf{e}_j)_e - (S_{o_h}, \nabla\cdot(N_i\mathbf{e}_j))_e + \langle \hat{S}_{o_h}, N_i\mathbf{e}_j\cdot\mathbf{n}\rangle_{\partial e}\right) \\
\left[\mathbf{R}_{\hat{S}}\right]_l &= \sum_{e\in \mathrm{T}_h} \left(\left\langle \mathbf{q}_{s_h}\cdot\mathbf{n} + \tau_s(S_{o_h} - \hat{S}_{o_h}) + \frac{\hat{\lambda}_o}{\hat{\lambda}_t}(\mathbf{q}_{p_h}\cdot\mathbf{n} + \tau_p(p_{w_h} - \hat{p}_{w_h})), N_l^f\right\rangle_{\partial e}\right) \\
&- \langle g_{Ns}^{\mathrm{out}}, N_l^f\rangle_{\Gamma_N^s}
\end{aligned}$$

3.3 Temporal Discretization

To integrate in time the DAE in Equation (24), we use a diagonally implicit Runge-Kutta method (DIRK). Specifically, we compute the oil saturation at time $t^{n+1} = t^n + \Delta t$ as

$$\mathbf{S}_o^{n+1} = \mathbf{S}_o^n + \Delta t \sum_{i=1}^{s} b_i \dot{\mathbf{S}}_o^{n,i},$$

where $\dot{\mathbf{S}}_o^{n,i}$ is the approximation of $\dot{\mathbf{S}}_o$ at time $t^{n,i}$. The oil saturation at each stage of the DIRK scheme is computed as

$$\mathbf{S}_o^{n,i} = \mathbf{S}_o^n + \Delta t \sum_{j=1}^{i} a_{ij} \dot{\mathbf{S}}_o^{n,j}, \tag{25}$$

and the $\dot{\mathbf{S}}_o^{n,i}$ for $i = 1, \ldots, s$ are computed as the solution of the non-linear algebraic equation:

$$\mathbf{R}\left(t^{n,i}, \mathbf{S}_o^n + \Delta t \sum_{j=1}^{i} a_{ij} \dot{\mathbf{S}}_o^{n,j}, \dot{\mathbf{S}}_o^{n,i}, \mathbf{q}_s^{n,i}, \hat{\mathbf{S}}_o^{n,i}, \mathbf{p}_w^{n,i}, \mathbf{q}_p^{n,i}, \hat{\mathbf{p}}_w^{n,i} \right) = \mathbf{0}. \tag{26}$$

Parameters b_i, c_i, a_{ij} define the DIRK method and are given by the Butcher's tables, see Table 1.

3.4 Non-linear Solver

To solve Equation (26), we use a fix-point iteration method. The main idea is to iteratively solve the saturation and the pressure until convergence is achieved. To this aim, we denote by l the l-th iteration of the non-linear solver. Thus, we first solve Equation (26) for the oil saturation by imposing

$$\begin{aligned} \mathbf{R}\Big(& \underbrace{t^{n,i}, \mathbf{S}_o^n + \Delta t \sum_{j=1}^{i-1} a_{ij} \dot{\mathbf{S}}_o^{n,j}}_{\text{Data}} + \underbrace{\Delta t a_{ii} \dot{\mathbf{S}}_o^{n,i,l+1}, \dot{\mathbf{S}}_o^{n,i,l+1}, \mathbf{q}_s^{n,i,l+1}, \hat{\mathbf{S}}_o^{n,i,l+1}}_{\text{Unknowns}}, \\ & \underbrace{\mathbf{p}_w^{n,i,l}, \mathbf{q}_p^{n,i,l}, \hat{\mathbf{p}}_w^{n,i,l}}_{\text{Data}} \Big) = \mathbf{0}, \end{aligned} \tag{27}$$

from which we compute $(\dot{\mathbf{S}}_o^{n,i,l+1}, \mathbf{q}_s^{n,i,l+1}, \hat{\mathbf{S}}_o^{n,i,l+1})$ given $(\mathbf{p}_w^{n,i,l}, \mathbf{q}_p^{n,i,l}, \hat{\mathbf{p}}_w^{n,i,l})$. Then, we compute $\mathbf{S}_o^{n,i,l+1}$ using Equation (25). Finally, we also solve Equation (26) for the water pressure by imposing

$$\begin{aligned} \mathbf{R}\Big(& \underbrace{t^{n,i}, \mathbf{S}_o^n + \Delta t \sum_{j=1}^{i-1} a_{ij} \dot{\mathbf{S}}_o^{n,j} + \Delta t a_{ii} \dot{\mathbf{S}}_o^{n,i,l+1}, \dot{\mathbf{S}}_o^{n,i,l+1}, \mathbf{q}_s^{n,i,l+1}, \hat{\mathbf{S}}_o^{n,i,l+1}}_{\text{Data}}, \\ & \underbrace{\mathbf{p}_w^{n,i,l+1}, \mathbf{q}_p^{n,i,l+1}, \hat{\mathbf{p}}_w^{n,i,l+1}}_{\text{Unknowns}} \Big) = \mathbf{0}, \end{aligned} \tag{28}$$

from which we obtain $(\mathbf{p}_w^{n,i,l+1}, \mathbf{q}_p^{n,i,l+1}, \hat{\mathbf{p}}_w^{n,i,l+1})$ given $(\dot{\mathbf{S}}_o^{n,i,l+1}, \mathbf{q}_s^{n,i,l+1}, \hat{\mathbf{S}}_o^{n,i,l+1})$.

This procedure is repeated until convergence is achieved at each Runge-Kutta stage, $i = 1, \ldots, s$. We define the stopping criteria of the non-linear solver as

$$
\begin{aligned}
&\frac{\|S_{oh}^{n,i,l} - S_{oh}^{n,i,l+1}\|_{L^2(\mathrm{T}_h)}}{\|S_{oh}^{n,i,l+1}\|_{L^2(\mathrm{T}_h)}} \le \varepsilon_{S_o}, &&\frac{\|p_{wh}^{n,i,l} - p_{wh}^{n,i,l+1}\|_{L^2(\mathrm{T}_h)}}{\|p_{wh}^{n,i,l+1}\|_{L^2(\mathrm{T}_h)}} \le \varepsilon_{p_w},\\
&\frac{\|\mathbf{q}_{sh}^{n,i,l} - \mathbf{q}_{sh}^{n,i,l+1}\|_{L^2(\mathrm{T}_h)}}{\|\mathbf{q}_{sh}^{n,i,l+1}\|_{L^2(\mathrm{T}_h)}} \le \varepsilon_{\mathbf{q}_s}, &&\frac{\|\mathbf{q}_{ph}^{n,i,l} - \mathbf{q}_{ph}^{n,i,l+1}\|_{L^2(\mathrm{T}_h)}}{\|\mathbf{q}_{ph}^{n,i,l+1}\|_{L^2(\mathrm{T}_h)}} \le \varepsilon_{\mathbf{q}_p},\\
&\frac{\|\hat{S}_{oh}^{n,i,l} - \hat{S}_{oh}^{n,i,l+1}\|_{L^2(\Sigma_h)}}{\|\hat{S}_{oh}^{n,i,l+1}\|_{L^2(\Sigma_h)}} \le \varepsilon_{\hat{S}_o}, &&\frac{\|\hat{p}_{wh}^{n,i,l} - \hat{p}_{wh}^{n,i,l+1}\|_{L^2(\Sigma_h)}}{\|\hat{p}_{wh}^{n,i,l+1}\|_{L^2(\Sigma_h)}} \le \varepsilon_{\hat{p}_w},\\
&\|\mathbf{R}_{S_o}\|_2 \le \varepsilon_{\mathbf{R}_{S_o}}, &&\|\mathbf{R}_{p_w}\|_2 \le \varepsilon_{\mathbf{R}_{p_w}},\\
&\|\mathbf{R}_{\mathbf{q}_s}\|_2 \le \varepsilon_{\mathbf{R}_{\mathbf{q}_s}}, &&\|\mathbf{R}_{\mathbf{q}_p}\|_2 \le \varepsilon_{\mathbf{R}_{\mathbf{q}_p}},\\
&\|\mathbf{R}_{\hat{S}_o}\|_2 \le \varepsilon_{\mathbf{R}_{\hat{S}_o}}, &&\|\mathbf{R}_{\hat{p}_w}\|_2 \le \varepsilon_{\mathbf{R}_{\hat{p}_w}}.
\end{aligned}
\tag{29}
$$

3.5 Local Post-processing

Similarly to one-phase flow, a local post-processing can be applied to obtain a new approximation for the pressure, p_h^*, and for the saturation, $S_{o_h}^*$, both in $\mathbb{V}_h^{P+1}$ with convergence rate of $P + 2$ in the L^2-norm, see [15]. For two-phase flow problem, we have two local problems. The first one consists of finding the post-processed pressure, $p_h^* \in \mathbb{V}_h^{P+1}$ on each element, e, such that:

$$
\left.\begin{aligned}
(\mathbf{K}\lambda_t \nabla p_h^*, \nabla v)_e &= -(\mathbf{q}_{p_h}, \nabla v)_e\\
(p_h^*, 1)_e &= (p_{w_h}, 1)_e
\end{aligned}\right\}
\tag{30}
$$

for all $v \in \mathbb{V}_h^{P+1}$. The second local problem consists of finding the post-processed saturation, $S_{o_h}^* \in \mathbb{V}_h^{P+1}$ on each element, e, such that:

$$
\left.\begin{aligned}
(\mathbf{K}\lambda_o p_c' \nabla S_{o_h}^*, \nabla v)_e &= -(\mathbf{q}_{s_h}, \nabla v)_e\\
(S_{o_h}^*, 1)_e &= (S_{o_h}, 1)_e
\end{aligned}\right\}
\tag{31}
$$

According to [15] it is important to highlight that this procedure can be applied at selected time steps, and it is not necessary to apply it at all the time steps [15, 18, 19].

Table 2 Butcher's table for a DIRK3-s3 scheme

$$\begin{array}{c|ccc} \gamma & \gamma & & \\ \frac{1+\gamma}{2} & \frac{1-\gamma}{2} & \gamma & \\ 1 & \frac{-6\gamma^2+16\gamma-1}{4} & \frac{6\gamma^2-20\gamma-1}{4} & \gamma \\ \hline & \frac{-6\gamma^2+16\gamma-1}{4} & \frac{6\gamma^2-20\gamma-1}{4} & \gamma \end{array}$$

4 Numerical Results

In this section, we present four examples of the proposed HDG formulations. The first two examples deal with one-phase flow, while the last two are related to two-phase flow. In all the examples, we use the diagonally implicit Runge-Kutta method of order three with three stages, DIRK3-s3, for the time discretization. The Runge-Kutta parameters are specified in Table 2, setting $\gamma = 0.4358665215$ [3].

The stopping tolerances of the Newton-Raphson method used in the two examples on one-phase flow are $\varepsilon_p = \varepsilon_{\mathbf{q}} = \varepsilon_{\hat{p}} = 10^{-10}$ and $\varepsilon_R = 10^{-5}$, see Equation (11). The stopping tolerances of the fix-point iterative procedure used in the two examples on two-phase flow are $\varepsilon_{S_o} = \varepsilon_{p_w} = \varepsilon_{\hat{S}_o} = \varepsilon_{\hat{p}_w} = 10^{-8}$, $\varepsilon_{\mathbf{q}_s}, = \varepsilon_{\mathbf{q}_p} = 10^{-6}$, $\varepsilon_{\mathbf{R}_{S_o}} = \varepsilon_{\mathbf{R}_{p_w}} = \varepsilon_{\mathbf{R}_{\hat{S}_o}} = \varepsilon_{\mathbf{R}_{\hat{p}_w}} = 10^{-8}$, and $\varepsilon_{\mathbf{R}_{\mathbf{q}_s}}, = \varepsilon_{\mathbf{R}_{\mathbf{q}_p}}, = 10^{-6}$, see Equation (29).

All the high-order meshes used in these examples are generated using the algorithms presented in [10, 11, 22] that are implemented in the EZ4U environment [21].

4.1 One-Phase Flow Through Homogeneous Material with an Obstacle

In this example, we present a 2D simulation of the flow generated by a well when an impermeable obstacle is located above it. We consider a square domain $\Omega = (0, 50) \times (0, 50)$ m with a circular well of radius $r_w = 1.7$ m located at the center of the domain, $\mathbf{x}_w = (25, 25)$ m, see Fig. 1. The parameters used for this example are defined in Table 3. We prescribe no-flow condition on all the boundaries, and the source term is

$$f^{2D} = \begin{cases} \dfrac{f}{\pi r_w{}^2} & \text{if } \sqrt{(x_w - x)^2 + (z_w - z)^2} < r_w, \\ 0 & \text{elsewhere.} \end{cases} \tag{32}$$

The domain, Ω, is discretized using an unstructured mesh of polynomial degree four with 96 quadrilateral elements (1632 nodes). The total number of unknowns is 8280. However, after applying the hybridization procedure, the size of the linear

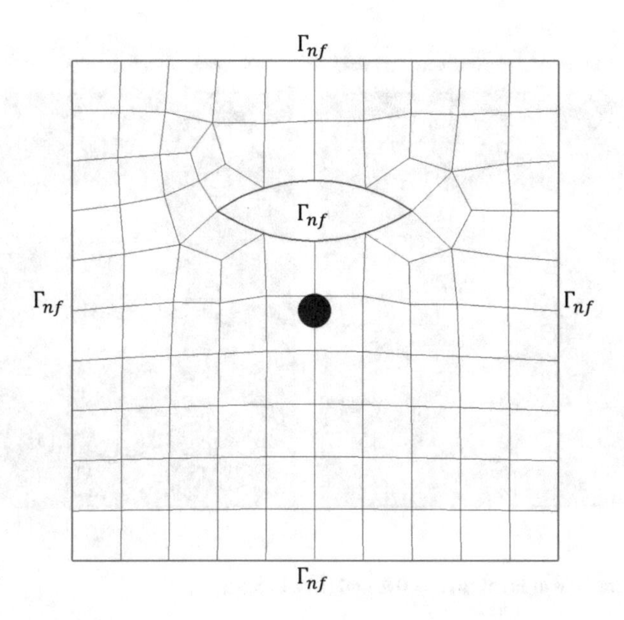

Fig. 1 Mesh and boundary conditions distributions for the obstacle one-phase example

Table 3 Material and fluid parameters for Example 4.1. (z is the reservoir depth)

Parameter	Parameter
$\mathbf{K} = 3 \cdot 10^{-14}$ m^2	$p^0 = 244.966 + 0.96z$ atm
$\phi_{\text{ref}} = 0.15$	$f = -0.25$ kg/s
$c_r = 5.8 \cdot 10^{-10}$ Pa^{-1}	$\mu = 0.001$ Pa s
$c_f = 1.45 \cdot 10^{-9}$ Pa^{-1}	$\rho_{\text{ref}} = 897.5$ kg/m^3

system to be solved is reduced to 1080. The time step used in this simulation is $\Delta t = 200$ s.

Note that we do not know a priori an initial condition compatible with the boundary condition, in which the hydrocarbon is totally still. To this end, we evolve the problem with a null source term until

$$\frac{\displaystyle\int_{\Omega} \|p^{n+1} - p^n\|^2 \mathrm{d}\Omega}{\displaystyle\int_{\Omega} 1 \mathrm{d}\Omega} < \varepsilon_{\text{abs}}, \tag{33}$$

where $\varepsilon_{\text{abs}} = 10^{-9}$. To perform this, we apply the Backward Euler scheme with a variable time step, $\Delta t_n = \Delta t_0 \cdot 1.105^n$, with $\Delta t_0 = 1.0$ s and n being the step number. Figure 2a shows the computed initial pressure distribution. Since we are only interested in the steady-state solution, we use the backward Euler scheme because it is unconditionally stable and large time steps can be used. Once the steady

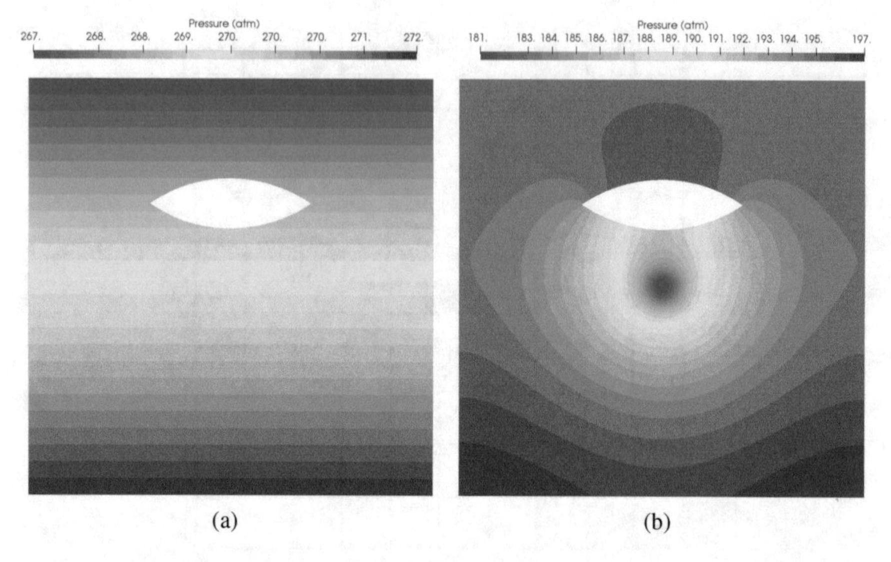

Fig. 2 Pressure field at time: (**a**) $t = 0\,\text{h}$, and (**b**) $t = 5.5\,\text{h}$.

state is obtained, we perform the time integration using a DIRK3-s3, because we are interested in an accurate tracking of the hydrocarbon extraction process.

Figure 2b shows the pressure approximation at time $t = 5.5\,\text{h}$. The pressure decreases near the well due to the hydrocarbon recovery, and it is higher at the bottom of the reservoir than the surface due to the gravity effect.

4.2 *One-Phase Flow Through Heterogeneous Material*

This example corresponds to a fully three-dimensional case with three different permeability regions, see Fig. 3a. The most permeable region is located at the middle, $\mathbf{K}_B$. At the bottom is the region with the lowest permeability, $\mathbf{K}_C$. The upper region has an intermediate permeability value, $\mathbf{K}_A$. The permeability values are detailed in Table 4.

The physical domain is $\Omega = (0, 50) \times (0, 50) \times (0, 50)$ m, and we impose no-flow condition on all the boundaries. We consider a spherical well with radius $r_w = 4.0$ m and the center located at $\mathbf{x}_w = (25, 25, 25)$ m. The source term is modeled as

$$f^{3D} = \begin{cases} \dfrac{f}{\dfrac{4}{3}\pi {r_w}^3} & \text{if } \sqrt{(x_w - x)^2 + (y_w - y)^2 + (z_w - z)^2} < r_w, \\ 0 & \text{elsewhere.} \end{cases}$$

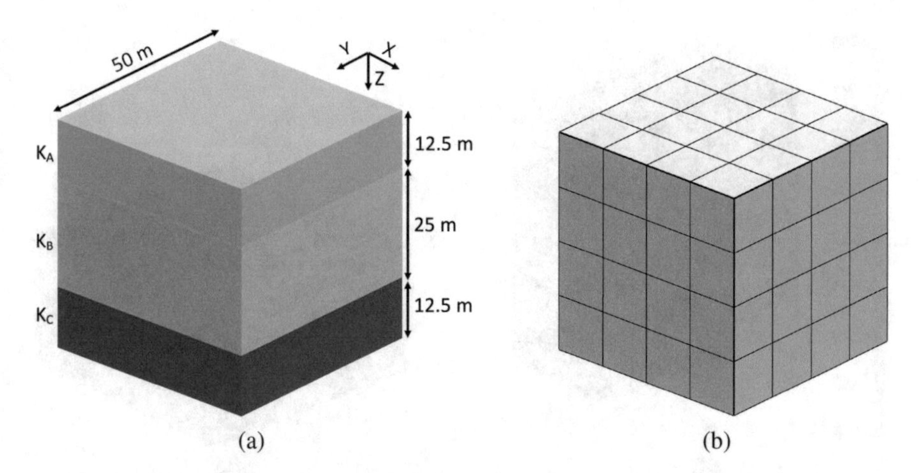

Fig. 3 (**a**) Permeability distribution. (**b**) Hexahedral elements of polynomial degree four and with an element size of 12.5 m

Table 4 Material and fluid parameters for Example 4.2

Parameter	Parameter
$\mathbf{K}_A = 10^{-14}$ m^2	
$\mathbf{K}_B = 10^{-13}$ m^2	$p^0 = 244.966 + 0.96z$ atm
$\mathbf{K}_C = 10^{-17}$ m^2	
$\phi_{\text{ref}} = 0.2$	$f = -2$ kg/s
$c_r = 5.8 \cdot 10^{-10}$ Pa^{-1}	$\mu = 0.00106$ Pa s
$c_f = 1.45 \cdot 10^{-9}$ Pa^{-1}	$\rho_{\text{ref}} = 897.5$ kg/m^3

The parameters used for this example are defined in Table 4. We discretize the domain, Ω, with a structured hexahedral mesh of 64 elements of polynomial degree four (4913 nodes), see Fig. 3b. The total number of unknowns is 38,000. However, after applying the hybridization procedure, the size of the linear system to be solved is reduced to 6000. The time step for this simulation is $\Delta t = 2$ h.

Similar to Example 4.1, first we have to compute an initial condition compatible with the boundary conditions in which the hydrocarbon is totally still. For that reason, we let the system evolve until Equation (33) is verified. To perform this, we apply the Backward Euler scheme with a variable time step. For this relaxation problem, we have used the same parameters than in Example 4.1.

Figure 4 shows the pressure field at time $t = 1$ day in two sections of the domain. The pressure increases with the depth due to the gravity effects, and it is lower near the source term, because of the hydrocarbon recovery. In addition, the pressure field in the impermeable zone remains higher than in the other two regions due to the low permeability value. In the other two regions the effect of the pumping well is negligible. Figure 4b shows a cross section perpendicular to the depth, in which the pressure field presents a circular symmetry centered at the well.

Figure 4a also shows that the Darcy velocity vectors are pointing to the well. Again, the different permeability regions affect to the Darcy velocity. In the middle

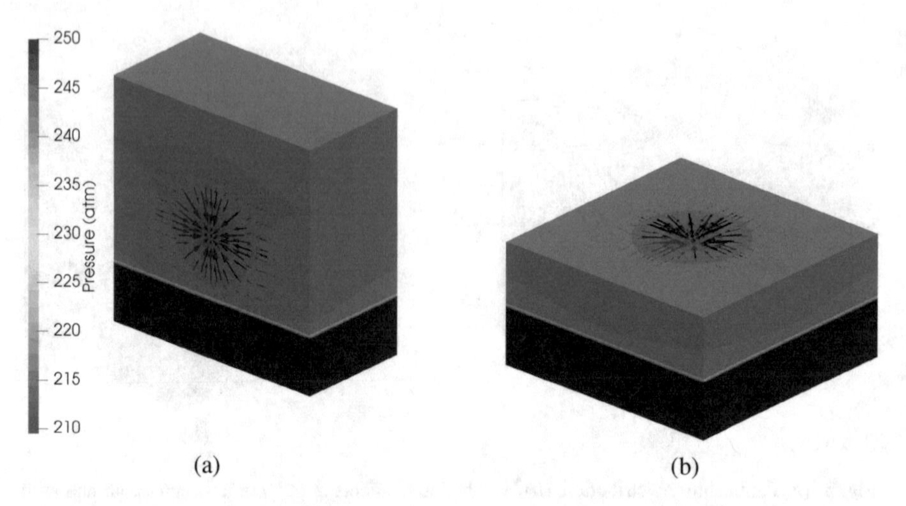

Fig. 4 Pressure field and Darcy's velocity in two sections of the computational domain: (**a**) YZ cross section, and (**b**) XY cross section

region, which has the highest permeability, $\mathbf{K}_B$, the fluid is moving faster than in the upper region, where the permeability is lower, $\mathbf{K}_A$, while the Darcy velocity is almost zero in the impermeable region, $\mathbf{K}_C$.

4.3 Water-Flooding Around Circular Obstacles

In this example, we simulate the water-flooding technique through a square domain, $\Omega = (0, 100) \times (0, 100)$ m, that contains five circular obstacles of radius $r_w = 5$ m located at (25, 25), (25, 50), (25, 75), (75, 37.5), (75, 62.5) m. The boundary is $\partial\Omega = \Gamma_{\text{in}} \bigcup \Gamma_{\text{out}} \bigcup \Gamma_{\text{nf}}$, see Fig. 5. We prescribe the boundary conditions as

$$\begin{aligned} p_w &= 3 \cdot 10^6 \text{ Pa}, & S_o &= 0.3, & &\text{on } \Gamma_{\text{in}}, \\ p_w &= 10^6 \text{ Pa}, & \left(\frac{\lambda_o \lambda_w}{\lambda_t} \mathbf{K} \nabla p_c\right) \cdot \mathbf{n} &= 0, & &\text{on } \Gamma_{\text{out}}, \\ \mathbf{v}_t \cdot \mathbf{n} &= 0, & \mathbf{v}_o \cdot \mathbf{n} &= 0, & &\text{on } \Gamma_{\text{nf}}. \end{aligned} \tag{34}$$

The soil permeability is $\mathbf{K} = 10^{-12}\mathbf{I}$ m^2, the porosity is $\phi = 0.2$, and the viscosity for the water and oil phases are $\mu_w = 0.001$ Pa · s and $\mu_o = 0.01$ Pa · s, respectively.

We discretize the domain using 433 quadrilateral elements of polynomial degree three (4058 nodes), see Fig. 5. The number of unknowns involved in the linear systems that have to be solved in each iteration of Equations (27) and (28) at each stage of the temporal scheme is 24468. Nevertheless, applying a static condensation procedure in the HDG formulation it is reduced to 3684 unknowns for each one.

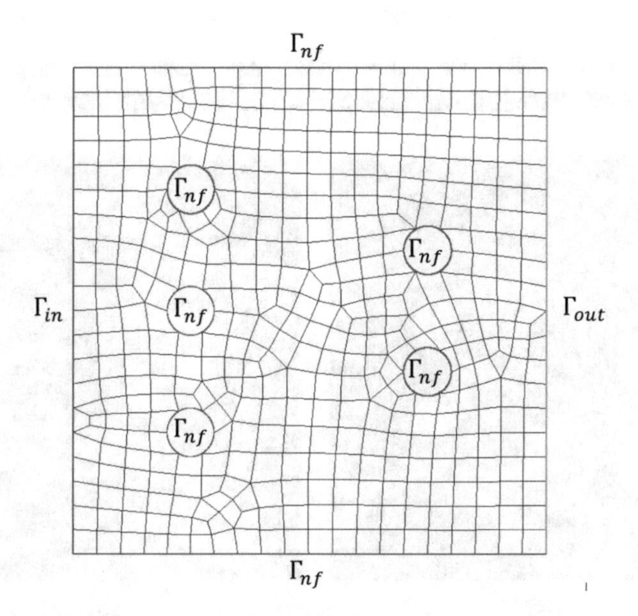

Fig. 5 Mesh and boundary conditions distributions for the circular boundaries example

The time step is $\Delta t = 1$ day. To mitigate the spurious oscillations generated by the sharp saturation fronts, we locally add artificial viscosity, see details in [20].

Figures 6 and 7 show the computed water saturation and water pressure at time $t = 20, 35, 50, 58$ days, respectively. Water is injected along the inflow boundary and moves the oil toward the outflow boundary, see Fig. 6. Note that water saturation has lower values behind each obstacle, see Fig. 6. In Fig. 7 we observe that the highest water pressure values are on the inflow boundary and the lowest on the outflow boundary. Furthermore, as expected at the left side of each circle the pressure is higher than at the right one, see Fig. 7.

Figure 8 shows the Darcy water velocity magnitude at time $t = 20, 35, 50, 58$ days. On the one hand, water velocity increases at the upper and bottom parts of each circular obstacle. On the other hand, the velocity is reduced at the left and right regions of each circular obstacle.

4.4 *Five-Spot Pattern*

In this example we consider a square domain, $\Omega = (0, 140) \times (0, 140)$ m. The selected pattern has four injection wells located at the vertices of square, and one producer well at its center, see Fig. 9. The radius of the wells is $r_w = 5$ m. On the boundary $\partial\Omega = \Gamma_{\text{in}} \bigcup \Gamma_{\text{out}} \bigcup \Gamma_{\text{nf}}$, we apply the boundary conditions detailed in Equation (34). The soil permeability is $\mathbf{K} = 5 \cdot 10^{-12}\mathbf{I}\ \text{m}^2$, the porosity is $\phi = 0.2$,

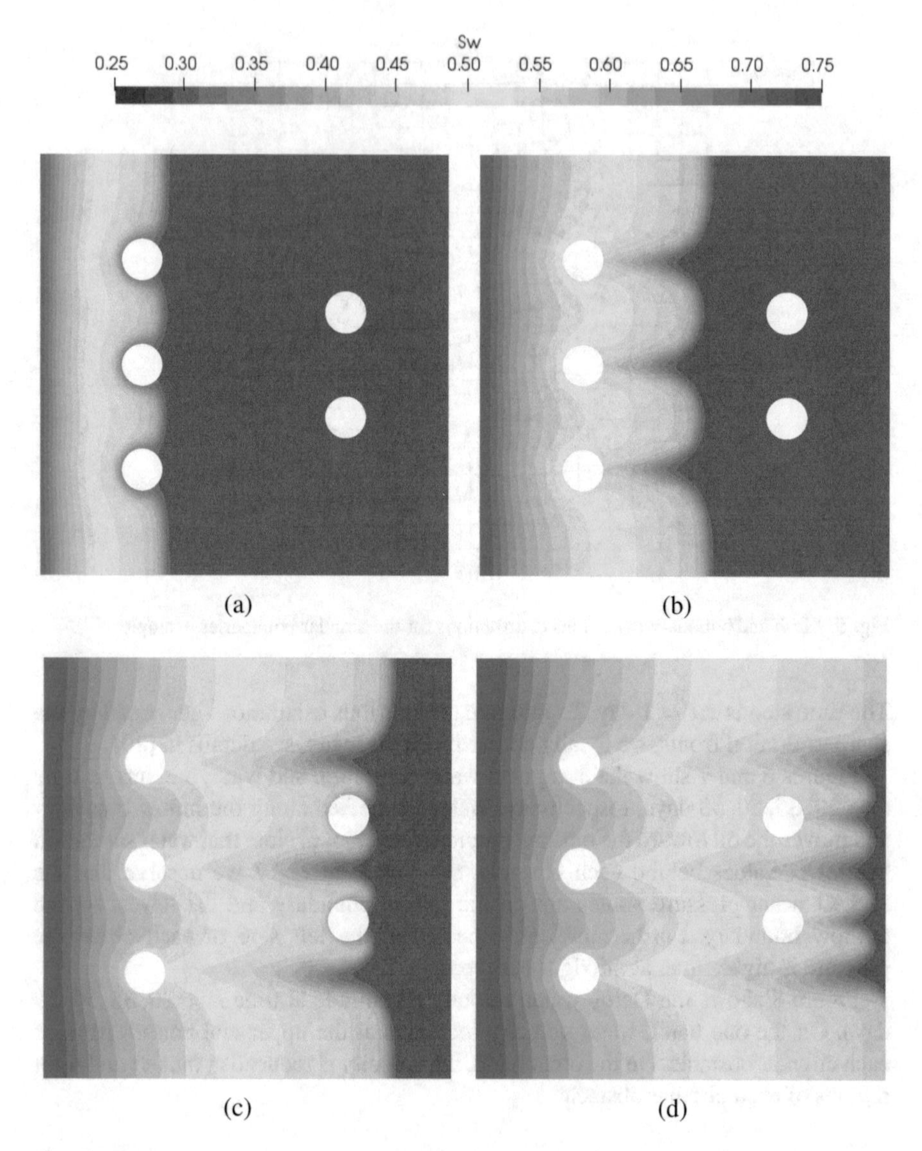

Fig. 6 Water saturation approximation at time: (**a**) 20 days, (**b**) 35 days, (**c**) 50 days, and (**d**) 55 days

and the viscosity for the water and oil phases are $\mu_w = 0.001$ Pa · s and $\mu_o = 0.012$ Pa · s, respectively.

We discretize Ω with 692 quadrilateral elements of polynomial degree four (11,372 nodes), see Fig. 9. The number of unknowns involved in the linear systems that have to be solved in each iteration of Equations (27) and (28) at each stage of the temporal scheme is 59,195. Nevertheless, applying a static condensation procedure

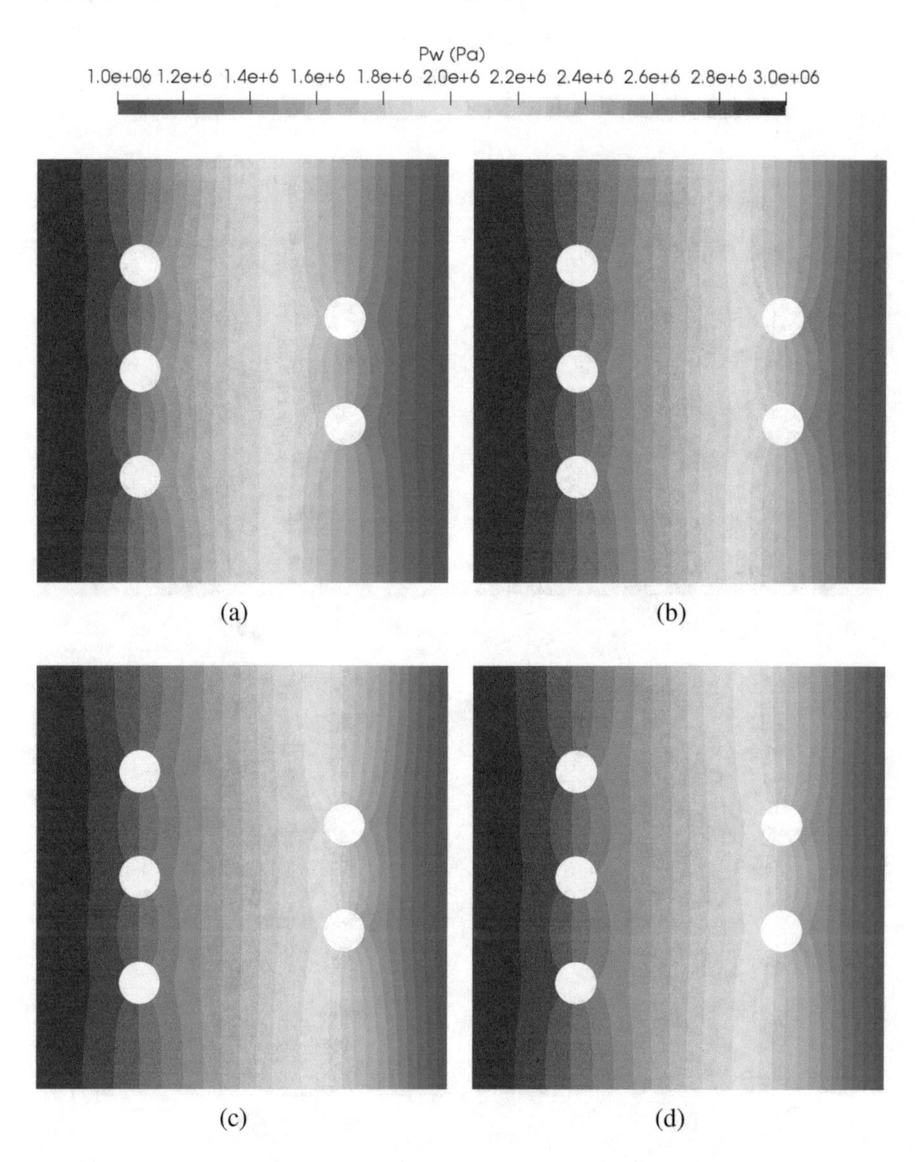

Fig. 7 Water pressure approximation at time: (**a**) 20 days, (**b**) 35 days, (**c**) 50 days, and (**d**) 55 days

in the HDG formulation it is reduced to 7295 unknowns for each one. The time step is $\Delta t = 12\,\text{h}$. To mitigate the spurious oscillations generated by the sharp saturation fronts, we locally add artificial viscosity, see details in [20].

Figure 10 shows the water saturation approximations at time $t = 5, 9, 11$ and 17 days. Water is injected from the corner wells (injectors), moving the oil to the pumping well at the center. Thus, water moves away from injector wells and

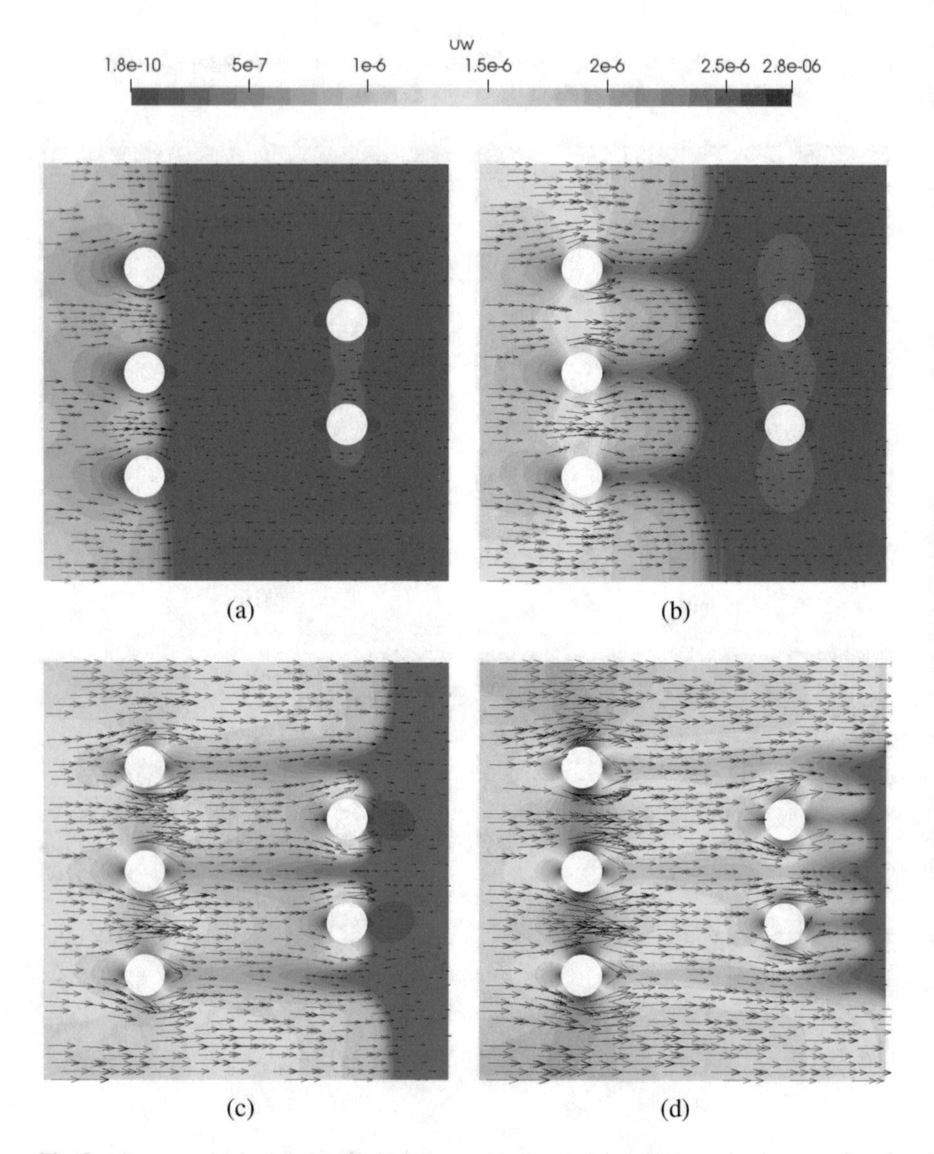

Fig. 8 Darcy water velocity magnitude approximation at time: (**a**) 20 days, (**b**) 35 days, (**c**) 50 days, and (**d**) 55 days

occupies the space left by the oil phase. Thus, the water saturation increases from the injectors wells to the producer well.

Figure 11 shows the water pressure field at time $t = 5, 9, 11$ and 17 days. As expected, the water pressure is higher at the injector wells than at the extractor well.

Figure 12 shows the water velocity and the oil velocity at time 9 and 17 days. We observe that the water velocity is higher around the injector, whereas the oil velocity

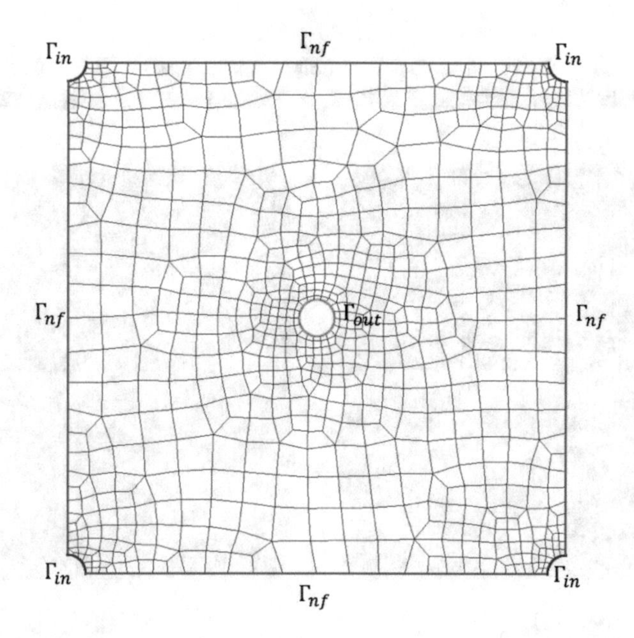

Fig. 9 Mesh and boundary conditions distributions for the five spot example

is higher around the extractor well, see Fig. 12. When the water front reaches the extractor well, both phases, oil and water, are extracted. At that time, water velocity increases and oil velocity decreases around the extractor well, see Fig. 12c, d.

5 Conclusions and Future Work

In this work, we have applied a high-order hybridizable discontinuous Galerkin formulation combined with high-order diagonally implicit Runge-Kutta temporal schemes for one-phase and two-phase flow problem through porous media. Specifically, the proposed formulation has been applied to solve four different examples dealing with 2D and 3D problems, and high-order structured and unstructured meshes. To this end, we have rewritten the initial governing equations as a set of first-order PDEs, and the weak form of each problems has been deduced.

For one-phase problems we have split the stabilization parameter into diffusive and convective parts. On the one hand, the diffusive one is computed in terms of the density and viscosity of the fluid, and the maximum eigenvalue of the permeability matrix. On the other hand, the convective one is evaluated using the Engquist-Osher monotone scheme flux. For two-phase problems, the stabilization parameters are computed in terms of the oil and total phase mobilities, and the maximum eigenvalue of the permeability matrix.

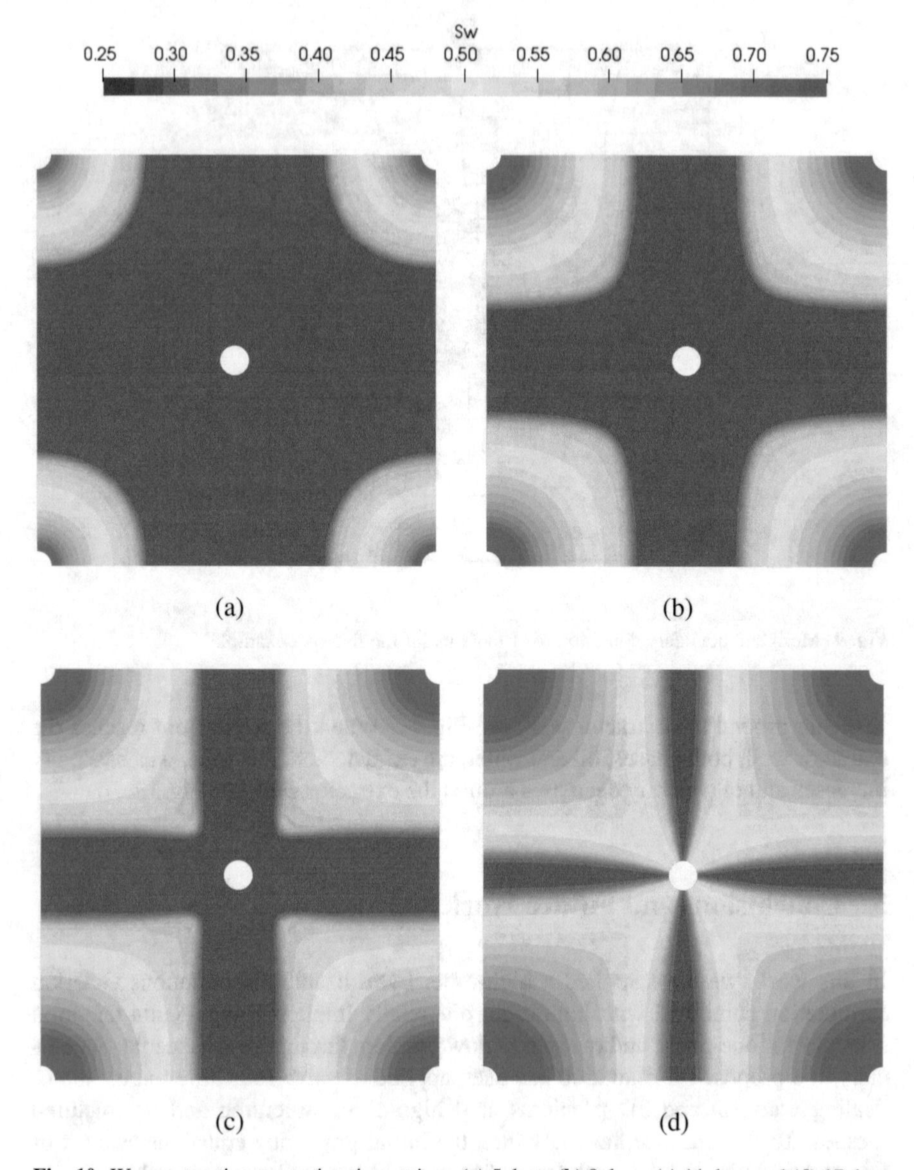

Fig. 10 Water saturation approximation at time: (**a**) 5 days, (**b**) 9 days, (**c**) 11 days and (**d**) 17 days

For one-phase problems we have used the Newton-Raphson method to solve the non-linear problem at each stage of the DIRK scheme. Typically, the number of iterations is around four if we consider homogeneous materials and two or three more for non-homogeneous materials. We found that the number of iterations depends on the reservoir heterogeneity.

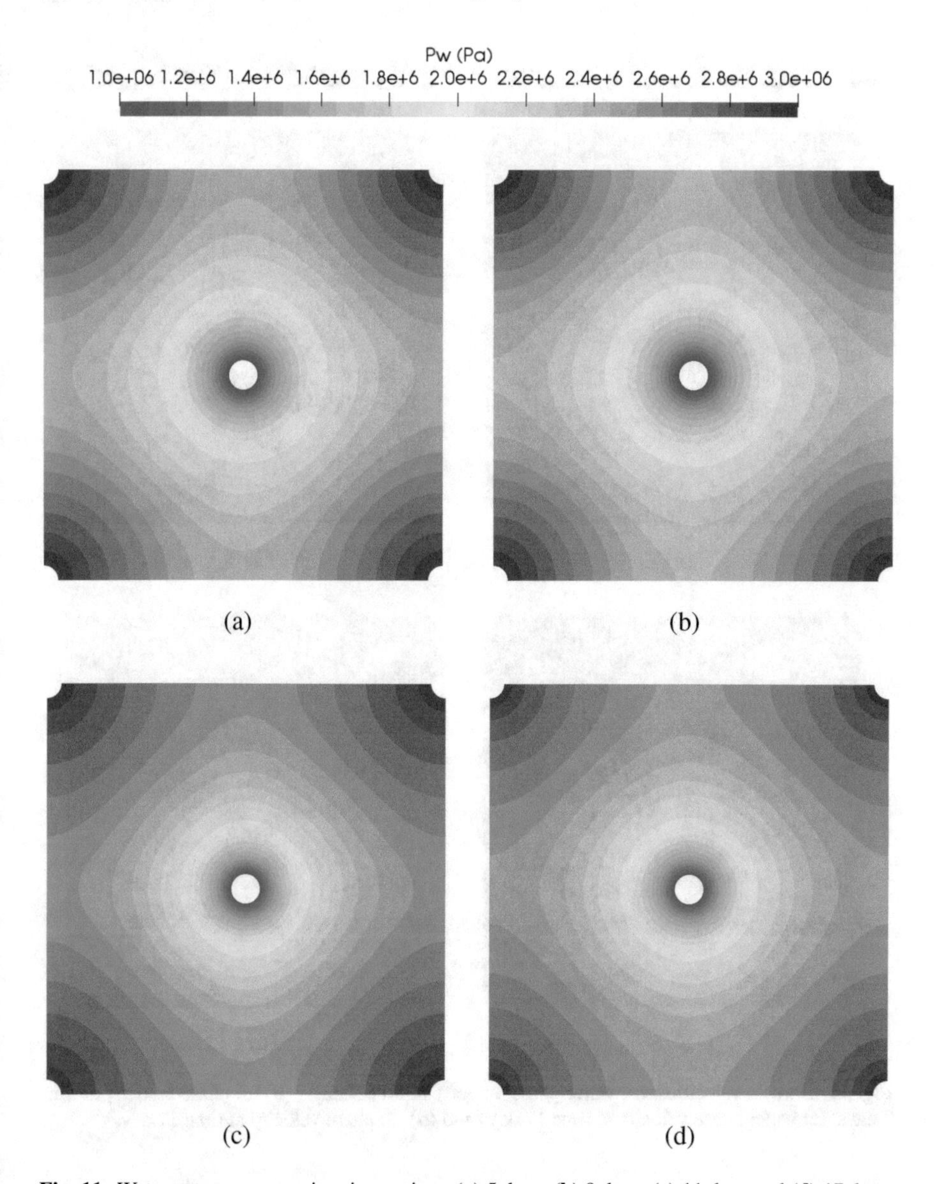

Fig. 11 Water pressure approximation at time: (**a**) 5 days, (**b**) 9 days, (**c**) 11 days and (**d**) 17 days

For two-phase flow problems, we have split the non-linear problem in two equations, the saturation and the pressure equation. We have used a fix-point iterative procedure to alternatively solve both variables until convergence is achieved at each stage of the DIRK scheme. This approach needs a larger number of iterations than the Newton-Raphson method. Nevertheless, the fix-point approach has smaller memory footprint than the Newton-Raphson method. We found that the number of

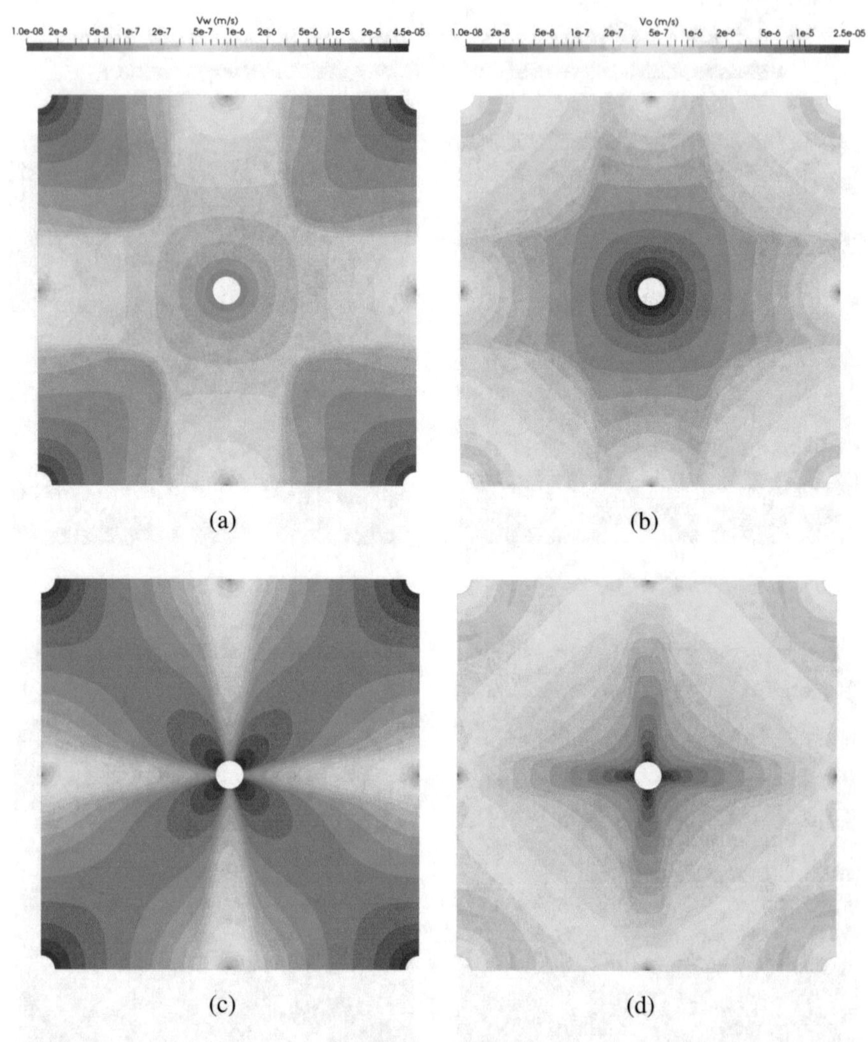

Fig. 12 Darcy's velocities: (**a**) water phase velocity at time 9 days, (**b**) oil phase velocity at time 9 days, (**c**) water phase velocity at time 17 days and (**d**) oil phase velocity at time 17 days

iterations of the fix-point procedure depends on the complexity of the domain, the size of the element, and the time step. For the two-phase flow examples presented in this paper the number of iterations is around thirty.

For both HDG formulations, we have used a Lagrangian basis of shape functions to define the elemental polynomial spaces and, therefore, the unknowns of the problem are the nodal values. Specifically, we have selected a non-uniform nodal distribution that approximately minimizes the Lebesgue constant, see [24]. Thus, the used basis is well-suited for high-order Lagrange interpolation.

Several aspects of this work will be analyzed and improved in the near future. First, we will investigate different higher-order temporal discretization schemes. Second, we will analyze a dimensionless formulation of both problems in order to improve the efficiency of our implementation. Third, we will explore alternative methods in order to improve the performance of the fix-point procedure used to solve the non-linear system at each stage of the DIRK scheme such as a Newton-Raphson method. Fourth, we will improve the computational efficiency of our implementation, which is currently programmed in Python, in order to apply our formulation to larger hydrocarbon reservoirs discretized with finer meshes. To this end, we will implement the proposed formulation using a compiled language like Fortran or C++. Moreover, we also consider to parallellize the code to further reduce the wall time.

Acknowledgments This work has been supported by FEDER and the Spanish Government, Ministerio de Economía y Competitividad grant project contract CTM2014-55014-C3-3-R, Ministerio de Ciencia Innovación y Universidades grant project contract PGC2018-097257-B-C33, and the grant BES-2015-072833.

References

1. Arbogast, T., Juntunen, M., Pool, J., Wheeler, M.F.: A discontinuous Galerkin method for two-phase flow in a porous medium enforcing H(div) velocity and continuous capillary pressure. Comput. Geosci. **17**(6), 1055–1078 (2013)
2. Bear, J., Verruijt, A.: Modeling Groundwater Flow and Pollution, vol. 2. Springer Science & Business Media, Netherlands (2012)
3. Butcher, J.C.: Numerical Methods for Ordinary Differential Equations. Wiley, England (2016)
4. Chen, Z., Huan, G., Ma, Y.: Computational Methods for Multiphase Flows in Porous Media, vol. 2. SIAM, Philadelphia (2006)
5. Corey, A.: Hydraulic properties of porous media. Colorado State University, Hydraulic Papers (3) (1964)
6. Costa-Solé, A., Ruiz-Gironés, E., Sarrate, J.: An HDG formulation for incompressible and immiscible two-phase porous media flow problems. Int. J. Comput. Fluid D. **33**(4), 137–148 (2019)
7. Donaldson, E., Chilingarian, G., Yen, T.: Enhanced oil recovery, II: processes and operations, vol. 17. Elsevier, Amsterdam/New York (1989)
8. Ern, A., Mozolevski, I., Schuh, L.: Discontinuous Galerkin approximation of two-phase flows in heterogeneous porous media with discontinuous capillary pressures. Comput. Methods Appl. Mech. Eng. **199**(23–24), 1491–1501 (2010)
9. Fabien, M.S., Knepley, M.G., Rivière, B.M.: A hybridizable discontinuous Galerkin method for two-phase flow in heterogeneous porous media. Int. J. Numer. Meth. Eng. **116**(3), 161–177 (2018)
10. Gargallo-Peiró, A., Roca, X., Peraire, J., Sarrate, J.: Optimization of a regularized distortion measure to generate curved high-order unstructured tetrahedral meshes. Int. J. Numer. Meth. Eng. **103**(5), 342–363 (2015)
11. Gargallo-Peiró, A., Roca, X., Peraire, J., Sarrate, J.: A distortion measure to validate and generate curved high-order meshes on CAD surfaces with independence of parameterization. Int. J. Numer. Meth. Eng. **106**(13), 1100–1130 (2016). Nme.5162

12. Jamei, M., Ghafouri, H.: A novel discontinuous Galerkin model for two-phase flow in porous media using an improved IMPES method. Int. J. Numer. Method H. **26**(1), 284–306 (2016)
13. Jamei, M., Raeisi Isa Abadi, A., Ahmadianfar, I.: A Lax-Wendroff-IMPES scheme for a two-phase flow in porous media using interior penalty discontinuous Galerkin method. Numer. Heat Tr. B.-Fund. **75**(5), 325–346 (2019)
14. Kennedy, C.A., Carpenter, M.H.: Diagonally implicit Runge-Kutta methods for ordinary differential equations a review. Technical Report NASA/TM-2016-219173, NASA (2016)
15. Kirby, R., Sherwin, S., Cockburn, B.: To CG or to HDG: a comparative study. J. Sci. Comput. **51**(1), 183–212 (2012)
16. Klieber, W., Rivière, B.: Adaptive simulations of two-phase flow by discontinuous Galerkin methods. Comput. Methods Appl. Mech. Eng. **196**(1–3), 404–419 (2006)
17. Montlaur, A., Fernandez-Mendez, S., Huerta, A.: High-order implicit time integration for unsteady incompressible flows. Int. J. Numer. Meth. Fl. **70**(5), 603–626 (2012)
18. Nguyen, N.C., Peraire, J., Cockburn, B.: An implicit high-order hybridizable discontinuous Galerkin method for linear convection–diffusion equations. J. Comput. Phys. **228**(9), 3232–3254 (2009)
19. Nguyen, N.C., Peraire, J., Cockburn, B.: An implicit high-order hybridizable discontinuous Galerkin method for nonlinear convection–diffusion equations. J. Comput. Phys. **228**(23), 8841–8855 (2009)
20. Persson, P.O., Peraire, J.: Sub-cell shock capturing for discontinuous Galerkin methods. In: 44th AIAA Aerospace Sciences Meeting and Exhibit, p. 112 (2006)
21. Roca, X., Ruiz-Gironés, E., Sarrate, J.: EZ4U: mesh generation environment (2010)
22. Ruiz-Gironés, E., Roca, X., Sarrate, J.: High-order mesh curving by distortion minimization with boundary nodes free to slide on a 3D CAD representation. Comput. Aided Des. **72**, 52–64 (2016)
23. Sevilla, R., Huerta, A.: Tutorial on hybridizable discontinuous Galerkin (HDG) for second-order elliptic problems. In: Advanced Finite Element Technologies, pp. 105–129. Springer, Cham (2016)
24. Warburton, T.: An explicit construction of interpolation nodes on the simplex. J. Eng. Math. **56**(3), 247–262 (2006)

Printed in the United States
by Baker & Taylor Publisher Services